Photographic Anatomy of the Human Body

PHOTOGRAPHIC ANATOMY OF THE HUMAN BODY

THIRD EDITION

Chihiro Yokochi, M.D.
Professor Emeritus
Department of Anatomy
Kanagawa Dental College
Yokosuka, Japan

Johannes W. Rohen, M.D.
Professor and Chairman
Department of Anatomy
University of Erlangen-Nürnberg
Erlangen, Germany

Eva Lurie Weinreb, Ph.D.
Professor, Department of Biology
Division of Life Sciences and Allied Health Services
Community College of Philadelphia
Philadelphia, Pennsylvania, U.S.A.

IGAKU-SHOIN Tokyo·New York

Published and distributed by

IGAKU-SHOIN Ltd.
 5-24-3 Hongo, Bunkyo-ku, Tokyo

IGAKU-SHOIN Medical Publishers, Inc.
 One Madison Avenue, New York, NY 10010

Library of Congress Cataloging-in-Publication Data

Yokochi, Chihiro.
    Photographic anatomy of the human body / Chihiro Yokochi, Johannes
 W. Rohen, Eva Lurie Weinreb. — 3rd ed.
         p.   cm.
    1. Anatomy, Human — Atlases.   I. Rohen, Johannes W. (Johannes
 Wilhelm)   II. Weinreb, Eva Lurie.   III. Title.
    [DNLM: 1. Anatomy — atlases.   QS 17 Y54j]
 QM25.Y613 1989
 611'.00222 — dc19
 DNLM/DLC   89-2047
 for Library of Congress   CIP

ISBN 4-260-14160-0 (Tokyo)
ISBN 0-89640-160-X (New York)

THIRD EDITION
Copyright © 1989 by IGAKU-SHOIN Ltd., Tokyo
Reprinted with revisions 1990, 1991,1992, 1994

10   9   8   7   6   5

Printed and bound in Japan

Preface

Gross anatomy remains one of the most important subjects in the study of medicine and allied health curricula. It is basic to all clinical studies. Although there are many anatomy books available throughout the world, it is hard to find a supplemental atlas that is suitable for both allied health students and professionals. Currently available atlases contain the traditional photographs and drawings of cadaver dissections. Because of the complexity of the dissections, it is difficult for students to have a clear concept of the three-dimensional structure of the body. The use of multiple, fully labeled, photographs accompanied by drawings helps students to understand the concepts of anatomy. Actual dissections of cadavers are still the best means of studying gross anatomy. However, this is often not possible because of the limitations of the physical facilities, the short time available for laboratory dissections, and the growing scarcity of suitable cadavers. It was with these needs in mind that the first edition of *PHOTO-GRAPHIC ANATOMY OF THE HUMAN BODY* by C. Yokochi was published by Igaku-Shoin Ltd. in 1969. Due to its worldwide acceptance, it was followed by the second edition by C. Yokochi and J. W. Rohen in 1978.

In the past decade we have witnessed the development of many new technologies for the study of anatomy and medical diagnosis, and for improved methods in color photography. In addition, we have seen an ever increasing demand for a suitable photographic atlas of anatomy that is geared to the needs of allied health students and to the orientation generally used in undergraduate courses in anatomy. With the publication of the third edition by C. Yokochi, J. W. Rohen and E. L. Weinreb, the authors have addressed the special needs of these students.

The third edition has been greatly expanded. It includes an introductory chapter to orient the student to the study of anatomy, many new photographs of dissections, new schematic drawings, and a comprehensive index. Some of the new photographs have also appeared in the second edition of the *COLOR ATLAS OF ANATOMY* by J.W. Rohen and C. Yokochi published by Igaku-Shoin and F.K. Schattauer in 1988. The preserved specimens were carefully chosen for their near normal appearance. In some cases fresh specimens were also used for contrast and to show certain anatomical details more clearly. Small and closely located structures, such as blood vessels, lymphatic vessels, and nerves, which may be difficult to distinguish are painted and color-coded for easy identification and tracing along their pathways. This technique was also used to clarify details of organ structure. The comprehensive index is especially helpful in an atlas where the same structures appear in so many different photographs.

This third edition was completely reorganized to make it most suitable for allied health students in the United States of America. It was designed to be a companion book to any textbook of anatomy or anatomy and physiology that may be used in undergraduate curricula.

The authors thank the many people, to whom they are indebted, who helped to make the publication of this atlas possible.

February 1989

Chihiro Yokochi, M.D.
Johannes W. Rohen, M.D.
Eva Lurie Weinreb, Ph.D.

Contents

1. Introduction to Anatomy

Anatomy is the study of the form and structure of the body and the relation between body parts. When this study is confined to body parts that are larger than 0.1 mm and visible to the naked eye, it is called gross anatomy. The origin of the term anatomy, from the Greek words *ana* and *tome* meaning to cut apart or dissect, reflects the method by which body structure is generally studied. Anatomical study may be approached by region or system. In **regional anatomy** the components of many systems such as bones, muscles, blood vessels, nerves, and viscera within a particular region such as the head or thorax are dissected. In **systemic anatomy** each body system is dissected completely and all component organs with their vascularization and innervation are examined. The systemic approach is often used in introductory courses and when functions are considered. The regional approach is preferred when clinical and surgical applications are considered. Because the cadavers used in the gross anatomy laboratory are generally dissected by region the regional approach is followed in this atlas.

Organization of the Body

The human body demonstrates bilateral symmetry. The right and left halves of the body are mirror images of each other. It is also segmented; however, segmentation is not as obvious in the adult as it is in the developmental stages. In the adult evidence of segmentation is clearly seen in the arrangement of vertebrae, blood vessels, and peripheral nerves. There are ten organ systems made up of multiple organs each composed of the four fundamental tissues (epithelia, connective, muscle, and nervous tissues). The systems are the integumentary, skeletal, muscular, nervous, circulatory (cardiovascular and lymphatic), respiratory, digestive, urinary, reproductive, and endocrine systems. Their component organs are as follows.

Integumentary system: skin and integumentary derivatives — hair, nails, sebaceous and sweat glands, part of mammary glands;

Skeletal system: bones, cartilages, articulations;

Muscular system: skeletal muscles and associated fibrous connective tissues—fascias, tendons, and aponeuroses;

Nervous system: brain and spinal cord in the central division and nerves, ganglia, and sensory receptors in the peripheral division;

Circulatory systems: heart and blood vessels — arteries, capillaries, and veins, in the cardiovascular system; lymphatic vessels and lymphoid tissues in the lymphatic system;

Respiratory system: nasal passages, pharynx, larynx, trachea, bronchial passages, and lungs;

Digestive system: mouth and accessory structures, pharynx, esophagus, stomach, small and large intestines, liver, biliary ducts, and exocrine pancreas;

Urinary system: kidneys, ureters, urinary bladder, and urethra;

Reproductive systems: sex glands (testes or ovaries), associated reproductive ducts and glands, and external genitalia;

Endocrine system: epiphysis (pineal gland), hypophysis (pituitary gland), thyroid, parathyroid, thymus, pancreatic islets, adrenal (suprarenal), and sex glands (testes, ovaries).

Organ systems may be located in more than one region. For example the digestive system begins in the head, continues through the neck, thorax, and abdomen, and terminates in the pelvis. Single organs may also occupy more than one region. For example the esophagus descends through the neck and thorax, penetrates the diaphragm, and enters the abdomen where it opens into the stomach. Components of the skeletal, muscular, vascular, and nervous systems are present in all body regions. To see the full extent of some structures such as arteries and nerves, which continue for long distances, it is necessary to identify these structures in dissections of different body regions that are shown in multiple photographs.

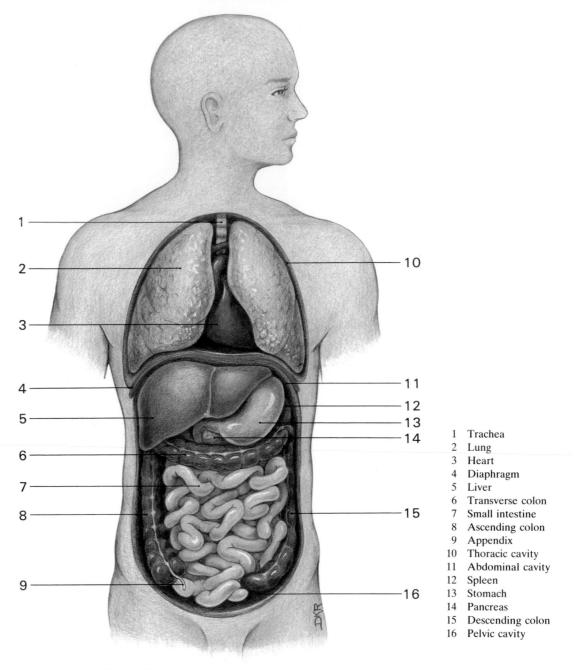

1	Trachea
2	Lung
3	Heart
4	Diaphragm
5	Liver
6	Transverse colon
7	Small intestine
8	Ascending colon
9	Appendix
10	Thoracic cavity
11	Abdominal cavity
12	Spleen
13	Stomach
14	Pancreas
15	Descending colon
16	Pelvic cavity

1.1   Ventral body cavity in anterior view showing thoracic and abdominopelvic organ contents.

Body Cavities

Internal organs are located within cavities of the body that are lined by smooth membranes. The **ventral cavity** towards the anterior surface is divided by the diaphragm into the thoracic cavity superiorly and the abdominal (abdominopelvic) cavity inferiorly (Fig. 1.1). The **thoracic cavity** contains the right and left pleural cavities, which hold the lungs, and the pericardial cavity, which holds the heart and proximal portions of the major blood vessels. Extending from the sternum anteriorly to the thoracic vertebrae posteriorly and from the thoracic inlet above to the diaphragm below are the organs of the thoracic cavity, exclusive of the lungs and pleurae, that constitute the mediastinum. These organs include the thymus, trachea, esophagus, heart and proximal blood

vessels, lymphatic vessels, lymph nodes, and nerves. The **abdominal cavity** that extends from the diaphragm to the pubis is arbitrarily divided into the abdominal cavity proper and pelvic cavity. The liver, bile ducts, spleen, pancreas, stomach, most of the small intestine, parts of the large intestine, kidneys, and ureters occupy the abdominal cavity proper. The sigmoid colon, rectum, urinary bladder, and parts of the male and female reproductive systems occupy the pelvic cavity. Because the pancreas, duodenum, rectum, kidneys, urinary bladder, and uterus (in females) are located against the posterior abdominal wall or pelvic floor and are not covered by peritoneum they are located retroperitoneally.

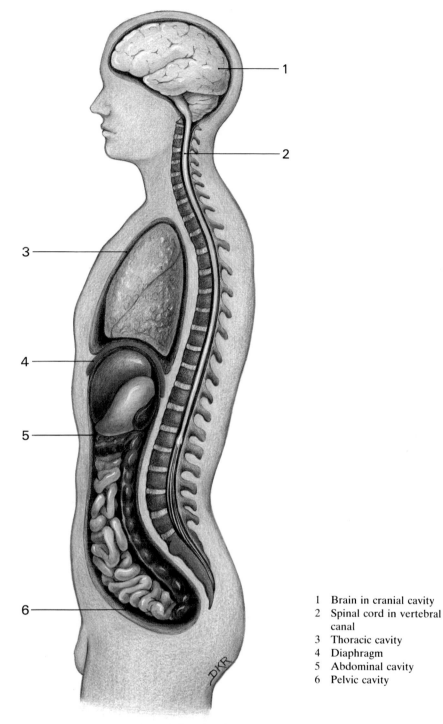

1 Brain in cranial cavity
2 Spinal cord in vertebral canal
3 Thoracic cavity
4 Diaphragm
5 Abdominal cavity
6 Pelvic cavity

1.2 Dorsal and ventral body cavities in lateral view showing organ contents.

The **dorsal cavity** is towards the posterior surface (Fig. 1.2). It consists of the **cranial cavity** in the skull, which contains the brain, and the **vertebral (spinal) canal** in the vertebral column, which contains the spinal cord and proximal portions of the nerve roots that emerge from the spinal cord and connect the spinal nerves with the spinal cord.

Anatomical Position and Directional Terms

To describe the location of body parts and the relationship between them the body is placed in the standard anatomical position (Fig. 1.3). In this position the body is in the upright posture, feet together flat on the ground, and arms hanging straight at the sides with the palms facing forward and thumbs directed away from the body. The location of body parts is described in terms of their relation to body regions including the head, front, back, midline, and sides of the body. The use of precise terminology precludes ambiguity and ensures accuracy in descriptions. Positional and directional terms are as follows (refer to Figs. 1.1 and 1.2).

Superior/Cranial means towards the head or uppermost part of the body; for example the shoulder is superior to the hip, but the hip is superior to the knee.
Inferior/Caudal means towards the tail end or lowermost part of the body; for example the hip is inferior to the shoulder, but the knee is inferior to the hip.
Anterior/Ventral means towards the front or belly surface of the body; for example the ribs and sternum of the thoracic cage are anterior to the lungs and heart, but the heart is anterior to the vertebral column.
Posterior/Dorsal means towards the back surface of the body opposite to the anterior surface; for example the vertebral column is posterior to the heart, but the heart is posterior to the sternum.

The terms ventral and dorsal are used in human anatomy with reference to body cavities and parts of the nervous system. These terms are routinely used to describe body parts towards the front or belly side and the back side, respectively, in animals that walk on all four legs.
Medial means towards the midline of the body; for example the sternum is medial.
Lateral means towards the sides of the body, away from the midline; for example the ribs are lateral to the sternum. The arms are described as lateral to the trunk of the body and the little finger as medial with the thumb as lateral in position.
Proximal means towards the place of attachment or origin of a body part or its relative closeness to its origin or attachment; for example the thigh bone, the femur, is proximal when compared to the leg bones, the tibia and fibula.
Distal means away from the place of attachment or origin of a body part or its relative distance from its origin or attachment; for example the tibia and fibula are distal when compared to the femur.
Superficial means closeness to the surface; for example the external oblique muscle is superficial compared to the internal oblique muscle.
Deep means below the surface at different levels; for example the internal oblique muscle is deep to the external oblique muscle. The terms **external** and **in-**

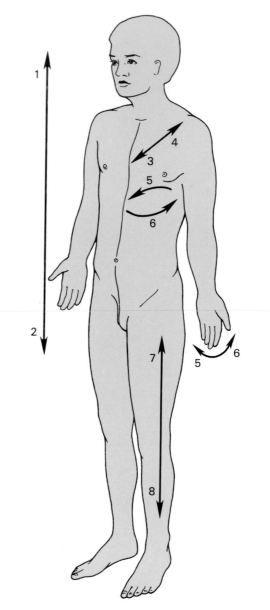

1.3 Commonly used positional and directional terms when the body is in standard anatomical position.

1	Superior/Cranial	5	Medial
2	Inferior/Caudal	6	Lateral
3	Anterior	7	Proximal
4	Posterior	8	Distal

ternal have meanings similar to those for superficial and deep, but are used to describe the locations of the body wall (external) and the viscera (internal).
Palmar refers to the anterior surface of the forearm and hand or palm.
Plantar refers to the sole of the foot.
The back of the hand and the top or instep of the foot are referred to as dorsal surfaces.

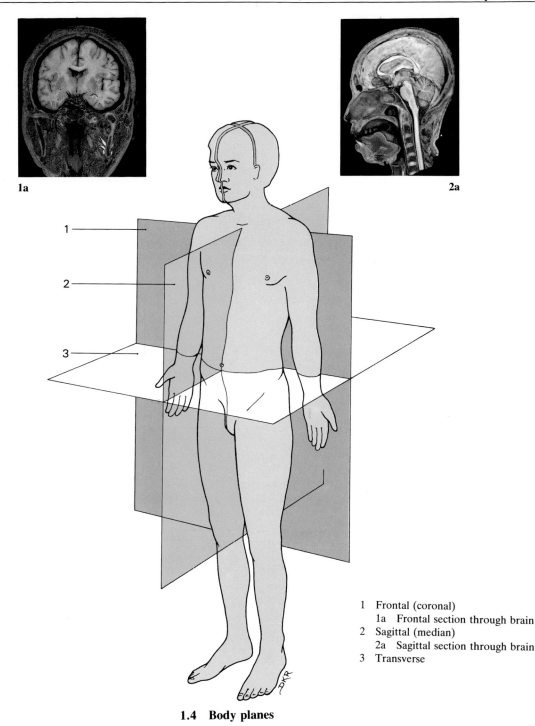

1a   2a

1   Frontal (coronal)
    1a   Frontal section through brain
2   Sagittal (median)
    2a   Sagittal section through brain
3   Transverse

1.4   Body planes

Body Planes

In dissection the body and body parts are cut along vertical or longitudinal planes and along horizontal or cross planes (Fig. 1.4). A vertical plane that divides the body into right and left portions is the **sagittal** or **median** plane. When it passes through the midline dividing the body into equal right and left halves, it is a midsagittal plane; when this plane passes to the right or left of the midline, it is a para-sagittal plane. A vertical plane made at right angles to the sagittal plane that divides the body into anterior and posterior portions is the **frontal** or **coronal** plane. A plane that passes at right angles to both the sagittal and frontal planes dividing the body into cross sections, which pass from anterior to posterior and right to left at any level, is the **transverse** plane. Sections through any level of the body, which show all structures present at a given level, may be visualized by means of radiological procedures such as computerized axial tomography (CT), positron emission tomography (PET), and magnetic resonance imaging (MRI).

Body Regions

To facilitate locating the internal organs the anterior body surface is divided into nine regions by imaginary lines, two parasagittal and two transverse lines (Fig.1.5). The regions are designated the right hypochondriac, epigastric, left hypochondriac, right lumbar, umbilical, left lumbar, right iliac or inguinal, hypogastric or pubic, and left iliac or inguinal regions. For less precise locations the body surface is divided into four quadrants by midsagittal and transverse lines that pass through the umbilicus. These are designated the upper right and left quadrants and lower right and left quadrants.

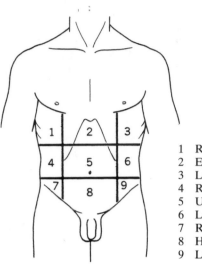

1 Right hypochondriac
2 Epigastric
3 Left hypochondriac
4 Right lumbar
5 Umbilical
6 Left lumbar
7 Right iliac (inguinal)
8 Hypogastric (pubic)
9 Left iliac (inguinal)

1.5 The nine regions of the abdomen (From Weinreb, E. L.: *Anatomy and Physiology*, Addison-Wesley Publishing Co., Reading, Mass., 1984).

Bones

A total of 206 bones compose the skeletal system. **Bones** are dynamic and metabolically active tissues that are involved in a variety of mechanical and metabolic functions. They support the body weight, give shape to parts of the body such as the head, protect internal organs such as the heart and lungs, and provide for attachment of muscles, tendons, and ligaments. Along with joints and muscles the bones form lever systems that enable different movements of the body. The bones are reservoirs for mineral salts such as calcium and phosphate salts and are sites for production of blood cells in the red bone marrow.

Bones are classified according to shape as short bones such as carpals; irregular bones such as vertebrae; flat bones such as the skull bones, sternum, and scapula; and long bones such as the femur (Figs. 1.6 and 1.7). Bone tissue is organized as cancellous (spongy) bone or compact (dense) bone. **Cancellous bone** is found in all bones. It forms the midportion of flat bones where it is sandwiched between layers of compact bone and fills the epiphyses of long bones. Cancellous bone consists of a latticework of bone trabeculae with red bone marrow filling the intervening spaces.

Compact bone forms the cortical portion of all bones. It forms the diaphysis of long bones where it encloses the medullary cavity. Compact bone consists of a precise arrangement of cylindrical structures called *Haversian systems* or osteons that are oriented parallel to the long axes of the bone (Fig. 1.8). Each Haversian system contains concentric lamellae of bone matrix around a central canal containing blood vessels, nerves and connective tissue. The lacunae, which hold the osteocytes, are evenly spaced between the lamellae. Canaliculi, into which cytoplasmic processes of the osteocytes extend, radiate outward from the lacunae to form an interconnecting network of channels throughout the bone matrix. Nutrients and metabolic products are exchanged between the osteocytes and matrix via the canaliculi.

Most of the outer surfaces of the bone, exclusive of those for articulation and attachment of tendons and ligaments, are covered by the tough connective tissue of the periosteum. The medullary cavity is lined by the finer connective tissue of the endosteum. Both periosteum and endosteum are sources of new bone cells. Periosteum is well vascularized and innervated and serves an important function in bone growth and repair. During the growth years a cartilaginous area, called the epiphyseal plate, separates the epiphyses from the diaphysis. This plate is joined to the diaphysis by the columns of cancellous bone in the metaphysis. When lengthwise growth and calcification of the bone have been completed, the plate is closed and only a remnant, called the epiphyseal line, remains.

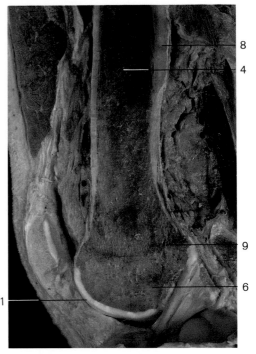

1.7 Sagittal section through distal end of femur of 20 year old male.

1   Articular cartilage
2   Cancellous bone with bone marrow
   a   Cancellous bone trabeculae
3   Compact bone
4   Medullary cavity with yellow bone marrow
   a   Medullary cavity
5   Periosteum
6   Epiphysis with red bone marrow
   a   Epiphysis
7   Site of endosteum
8   Diaphysis
9   Epiphyseal line

1.6 Femur of the adult. Bone on left shows periosteum and blood vessels over the diaphysis, and articular cartilage over the distal epiphysis. Bone on right is sectioned to show the trabeculae in cancellous bone of the epiphyses, and the medullary cavity and compact bone of the diaphysis.

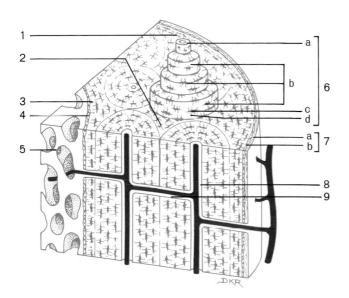

1.8 Schematic diagram of **compact bone** showing arrangement of Haversian systems in cross and longitudinal sections.

1   External circumferential lamellae
2   Interstitial lamellae
3   Internal circumferential lamellae
4   Endosteum
5   Trabeculae of cancellous bone
6   Haversian system
   a   Central canal
   b   Lamellae
   c   Lacunae with osteocytes
   d   Canaliculi
7   Periosteum
   a   Outer fibrous layer
   b   Inner osteogenic layer
8   Blood vessels and endosteal lining of Haversian canal
9   Volkmann's canal

Articulations (Joints)

The functional connections between bones are the **articulations** or **joints**. They are classified according to structure or function (Table 1.1). Fibrous and cartilaginous joints lack a joint cavity between articulating bones. *Fibrous joints* such as sutures are held tightly by fibrous connective tissue, are practically immovable, and termed synarthroses. *Cartilaginous joints* such as symphyses are held by fibrocartilage, have only slight movement, and are termed amphiarthroses. The majority of the joints in the body are the synovial joints that are freely movable and termed diarthroses. Synovial joints have a joint cavity that is filled with synovial fluid and enclosed by a joint capsule (Fig.1.9). The capsule is formed by the tough outer fibrous capsule continuous with the periosteum and the inner synovial membrane. Articulating surfaces of the bones are covered by hyaline cartilage. In some joints there are additional synovial tendon sheaths or synovial bursae that ease the movement of tendons and muscles over the bony processes.

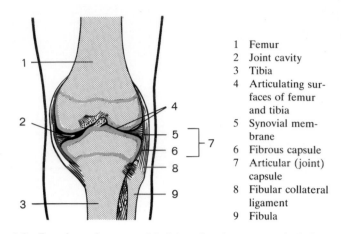

1 Femur
2 Joint cavity
3 Tibia
4 Articulating surfaces of femur and tibia
5 Synovial membrane
6 Fibrous capsule
7 Articular (joint) capsule
8 Fibular collateral ligament
9 Fibula

1.9 Drawing of a **synovial joint** showing two articulating bones and the joint cavity in frontal (coronal) section through the knee joint.

Table 1.1   Classification of Joints

	Type	Movement	Examples
A.	Fibrous joints- without a joint cavity	synarthroses (practically immovable)	
	1. Suture		between skull bones (frontoparietal, parieto-temporal)
	2. Syndesmosis		distal tibiofibular joint
	3. Gomphosis		teeth in alveolar processes of maxilla and mandible
B.	Cartilaginous joints-without a joint cavity	amphiarthroses (slightly movable)	
	1. Symphysis		between pubic bones of os coxae (pubic symphysis)
	2. Synchondrosis		between diaphysis and proximal or distal epiphysis of long bones
C.	Synovial joints- with a joint cavity	diarthroses (freely movable)	
	1. Hinge/ Ginglymus	monaxial (flexion-extension)	elbow (humeroulnar), interphalangeal
	2. Pivot/ Trochoid	monaxial (rotation, supination-pronation)	between vertebrae C1 and C2 (atlanto-axial), proximal radioulnar
	3. Gliding/ Arthrodia	monaxial	intercarpal, intertarsal, between ribs and thoracic vertebrae
	4. Condyloid/ Ellipsoidal	biaxial (flexion-extension, adduction-abduction)	between skull and cervical vertebra C1 (atlanto-occipital), radio-carpal
	5. Saddle	same as condyloid, but freer	carpo-metacarpal of thumb
	6. Ball and Socket	triaxial (flexion-extension, adduction-abduction, rotation)	shoulder (humerus and glenoid cavity of scapula), hip (femur and acetabulum of os coxa)

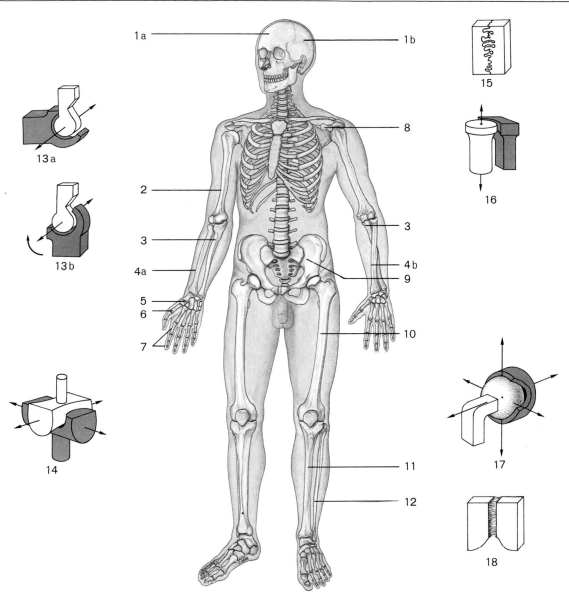

1.10 Joints of the axial and appendicular skeleton.

Bones

1	Skull
	a Frontal
	b Parietal
2	Humerus
3	Ulna
4a	Radius (supinated)
4b	Radius (pronated)
5	Carpal (trapezium)
6	Metacarpal I
7	Phalanges
8	Scapula
9	Os coxa
10	Femur
11	Tibia
12	Fibula

Joints (arrows: axes of movement)

13 Hinge (humeroulnar)
    a Flexion
    b Extension
14 Saddle (carpo-metacarpal)
15 Suture (frontoparietal)
16 Pivot (proximal radioulnar)
17 Ball and socket
    (humerus and scapula, femur and os coxa)
18 Syndesmosis (distal tibiofibular)

Synovial joints are classified according to shape and movement (Fig. 1.10). The movements are angular such as flexion and extension at the elbow, and adduction and abduction at the shoulder; rotation such as supination and pronation of the forearm; circumduction in which a complete circle is described by a bone such as the humerus in the glenoid cavity of the scapula; and gliding of one surface over another such as between the ribs and thoracic vertebrae. The classification according to movement includes the hinge joint that allows movement in one plane (monaxial), saddle and condyloid joints that allow movement in two planes (biaxial), pivot joints that are monaxial and allow rotation, and ball and socket joints that allow movement around several axes (triaxial).

Muscles

Muscle tissue is classified as striated and nonstriated or smooth. Striated muscles include those attached to the bones and back of the eyeballs, in the tongue, and the myocardium of the heart. To distinguish between striated muscles, heart muscle is referred to as cardiac muscle and other striated muscles are generally called skeletal muscles. Skeletal muscles are innervated by general somatic motor neurons and produce voluntary movements. Cardiac and smooth muscles are innervated by visceral (autonomic) motor neurons and produce involuntary movements. Smooth muscle is found in the organs of the respiratory, digestive, urinary, and reproductive systems and in the walls of ducts and blood vessels.

The muscular system is composed of more than 700 skeletal muscles and associated connective tissues and accounts for about 40% of the body weight. **Skeletal muscle** is composed of cells (myofibers) arranged in bundles (fascicles) supported by fibrous connective tissues (Fig. 1.11). Muscle is covered externally by the deep fascia that forms the epimysium and extends inward to form the perimysium supporting the fascicles. Connective tissue fibers extend beyond the myofibers to form narrow tendons or broad aponeuroses by which the muscle is attached to bones or other muscles. A muscle is attached at two points—the *origin*, which is the more fixed end, and the *insertion*, which is the more movable end. The portion between origin and insertion, called the belly or gaster, generally overlies the surface of the bone.

The movement produced by a muscle is called its action. It is due to muscles working in pairs to produce opposing actions. For example, flexion of the forearm is due to contraction of the biceps brachii on the anterior surface of the humerus and relaxation of the triceps brachii on the posterior surface of the humerus. Muscles are named by different criteria. These include the direction of fascicles such as straight (rectus), across (transverse), or diagonal (oblique); location such as overlying bone (frontalis) or extending between bones (intercostal); relative size such as small (minimus), large (maximus), or short (brevis); shape such as triangular (deltoid) or long and round (teres); number of points of attachment such as two heads (biceps) or three heads (triceps); type of movement such as flexion (flexors) or extension (extensors); and place of origin and insertion (such as the sternocleidomastoid).

Muscle fascicles and their tendons are arranged in different patterns which reflect their movements (Fig. 1.11). These arrangements are categorized as longitudinal (parallel or fusiform), radiate (convergent), pinnate (pennate), and circular (ringlike). In **longitudinal muscles** the fascicles are parallel with the longitudinal axis and the muscle is straplike, such as the sartorius and rectus abdominis. In **radiate muscles** the fascicles converge from a broad end to a narrower end, such as the pectoralis, deltoid, and latissimus dorsi. In **pinnate muscles** the fascicles run obliquely and converge like the plumes of a feather into the side of the tendon of insertion. When fascicles converge to one side, such as the semimembranosus, they are termed *unipennate*; when they converge to both sides of the tendon, such as the tibialis anterior, they are *bipennate*; and when there are multiple tendons of insertion, such as the deltoid, they are *multipennate*. In **circular muscles** the fascicles enclose an orifice and function as a sphincter, such as the orbicularis oculi. Muscles with multiple origins (heads) are described as bicipital, tricipital, and quadricipital such as the biceps brachii, triceps brachii or triceps surae (gastrocnemius and soleus), and quadriceps femoris, respectively.

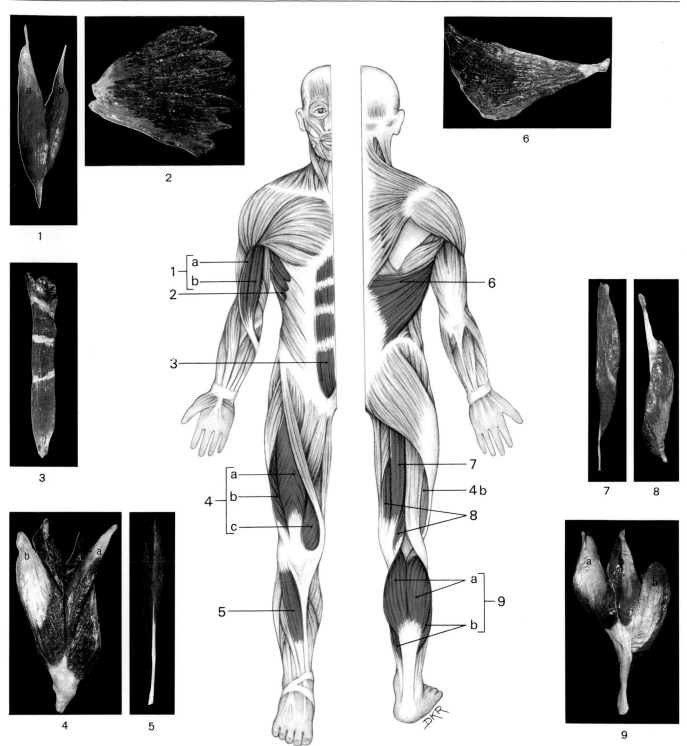

1.11 **The muscular system** in anterior (left) and posterior (right) views, with different muscle arrangements shown in inserts.

1	Biceps brachii	4	Quadriceps femoris	5	Tibialis anterior
	a Long head		a Rectus femoris	6	Latissimus dorsi
	b Short head		b Vastus lateralis	7	Semitendinosus
2	Serratus anterior		c Vastus medialis	8	Semimembranosus
3	Rectus abdominis		d Vastus intermedius		

9 Triceps surae
   a Gastrocnemius
    (medial and lateral heads)
   b Soleus

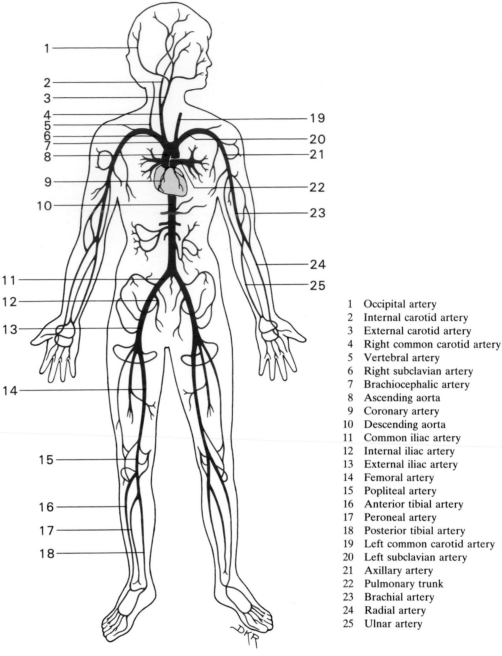

1   Occipital artery
2   Internal carotid artery
3   External carotid artery
4   Right common carotid artery
5   Vertebral artery
6   Right subclavian artery
7   Brachiocephalic artery
8   Ascending aorta
9   Coronary artery
10   Descending aorta
11   Common iliac artery
12   Internal iliac artery
13   External iliac artery
14   Femoral artery
15   Popliteal artery
16   Anterior tibial artery
17   Peroneal artery
18   Posterior tibial artery
19   Left common carotid artery
20   Left subclavian artery
21   Axillary artery
22   Pulmonary trunk
23   Brachial artery
24   Radial artery
25   Ulnar artery

1.12   The arterial system

Blood Vessels

The circulatory systems of the body are the cardiovascular and lymphatic systems. They are the transport systems that provide the anatomical and physiological links between body parts. The cardiovascular system composed of the heart and blood vessels transports blood through two main circulatory routes. The pulmonary route is between the heart and lungs; the systemic route is between the heart and rest of the body. The **blood vessels** form a closed system consisting of arteries, capillaries, and veins. The **arterial system** (Fig. 1.12) begins with the main trunk artery, the aorta, which leaves the left ventricle of the heart. The aorta gives off the branch arteries which, in turn, divide into arterial vessels of progressively smaller caliber that deliver blood to the tissues. Large arteries, such as the aorta and pulmonary arteries, are classified as elastic arteries because of the elastic tissue in their walls. Medium arteries, such as the brachial and radial arteries, are classified as muscular arteries because of the prominent smooth muscle layer in their walls. The smallest arteries are the arterioles. Their muscular walls provide the peripheral resistance which, under autonomic and endocrine regulation, maintains the pressure in the arterial system. Arterioles branch into the smallest vessels, the capillaries, which form extensive networks in the microcirculation. Here the exchange of oxygen, nutrients, and metabolic products takes place between the body fluids and cells.

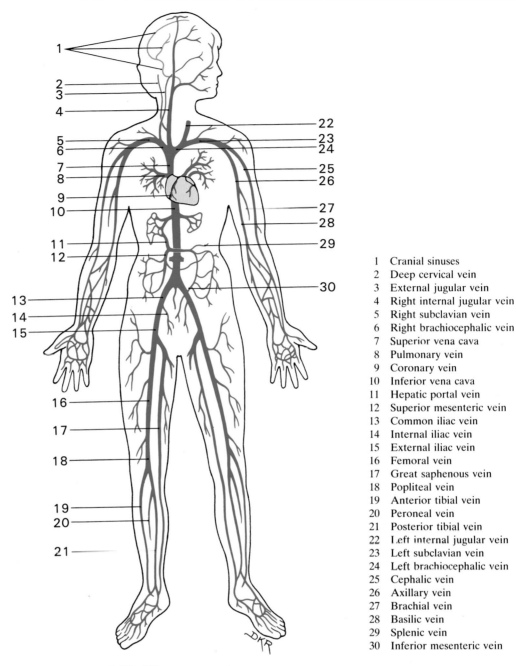

1   Cranial sinuses
2   Deep cervical vein
3   External jugular vein
4   Right internal jugular vein
5   Right subclavian vein
6   Right brachiocephalic vein
7   Superior vena cava
8   Pulmonary vein
9   Coronary vein
10 Inferior vena cava
11 Hepatic portal vein
12 Superior mesenteric vein
13 Common iliac vein
14 Internal iliac vein
15 External iliac vein
16 Femoral vein
17 Great saphenous vein
18 Popliteal vein
19 Anterior tibial vein
20 Peroneal vein
21 Posterior tibial vein
22 Left internal jugular vein
23 Left subclavian vein
24 Left brachiocephalic vein
25 Cephalic vein
26 Axillary vein
27 Brachial vein
28 Basilic vein
29 Splenic vein
30 Inferior mesenteric vein

1.13   The venous system

The **venous system** (Fig. 1.13) begins with the smallest veins, called venules, which leave the capillaries. Venules converge to form the tributary veins of progressively larger caliber that return blood to the heart. Veins are superficial or deep. Deep veins generally follow the course of corresponding arteries and may bear the same names. Blood is returned to the right atrium of the heart from the head, neck, thorax, and upper extremities via the superior vena cava; from the abdomen, pelvis, and lower extremities via the inferior vena cava; and from most coronary veins via the coronary sinus. Some tributary veins exhibit variable arrangements; for example veins form plexuses, such as the pelvic venous plexus, or sinuses, such as the large sagittal sinuses into which veins of the dura mater drain and the smaller sinuses that are found in the liver, spleen, and hypophysis. In superficial veins, where there is low venous pressure, valves are present in the intimal layer that help direct the flow of blood forward towards the heart and prevent backflow.

The distribution of the major blood vessels follows that of the peripheral nerves. Blood vessels and nerves that supply the same tissues and organs often bear the same names; for example, axillary, radial, and ulnar arteries are named the same as the corresponding nerves. Branch arteries are named for the tissues and organs they supply or through which they pass. The renal arteries are named for the renal (kidney) tissues they supply, whereas the subclavian, axillary, and brachial arteries are named for the regions of the shoulder and arm through which they pass before giving off branch arteries to the forearm and hand.

2. The Human Skeleton

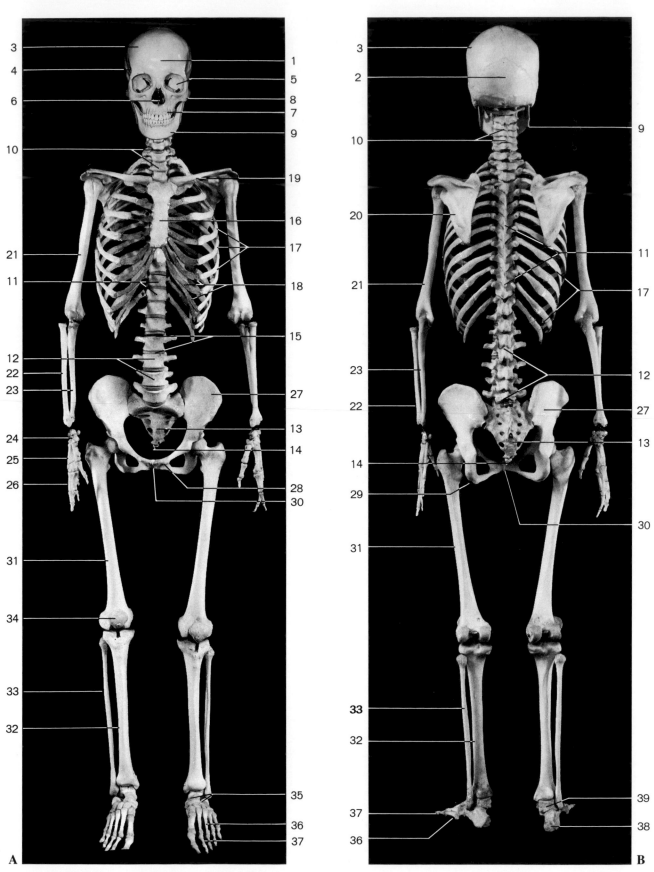

2.1 **Skeleton of the adult** (female), anterior (A) and posterior (B) views.

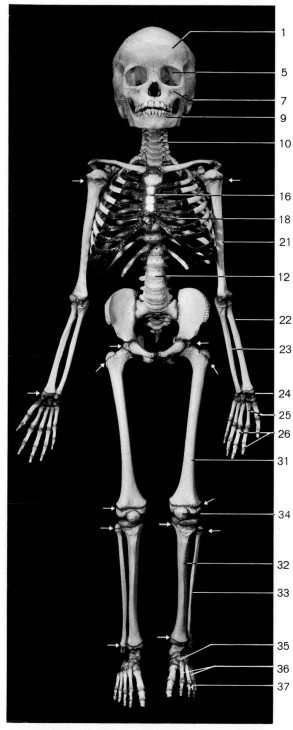

2.2 Skeleton of a 5 year old child (anterior view).
In contrast to the adult the child's skeleton shows cartilaginous growth (epiphyseal) plates (arrows); ilium, ischium and pubis of the hip bone are joined by cartilage; and three parts of the sternum are separated by cartilage. The ribs are predominantly horizontal in position.

Head
1   Frontal bone
2   Occipital bone
3   Parietal bone
4   Temporal bone
5   Orbit
6   Nasal cavity
7   Maxilla
8   Zygomatic bone
9   Mandible

Trunk and Thorax
Vertebral column
10   Cervical vertebrae
11   Thoracic vertebrae
12   Lumbar vertebrae
13   Sacrum
14   Coccyx
15   Intervertebral disks
Thorax
16   Sternum
17   Ribs
18   Costal cartilage

Upper limb and Pectoral (shoulder) girdle
19   Clavicle
20   Scapula
21   Humerus
22   Radius
23   Ulna
24   Carpal bones
25   Metacarpal bones
26   Phalanges of hand

Lower limb and Pelvic (hip) girdle
27   Ilium
28   Pubis
29   Ischium
30   Pubic symphysis
31   Femur
32   Tibia
33   Fibula
34   Patella
35   Tarsal bones
36   Metatarsal bones
37   Phalanges of foot
38   Calcaneus
39   Talus

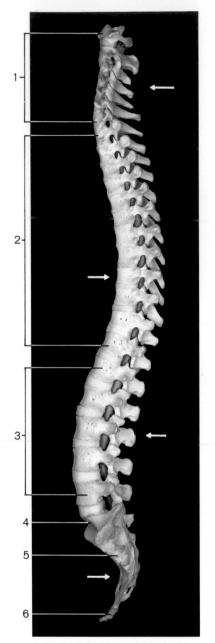

2.3   Vertebral column (left lateral view).   Arrows indicate cervical, thoracic, lumbar and sacral curves.

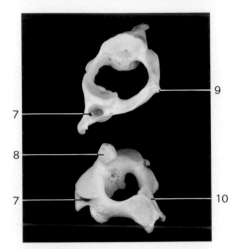

2.4   Atlas and axis (superior view).

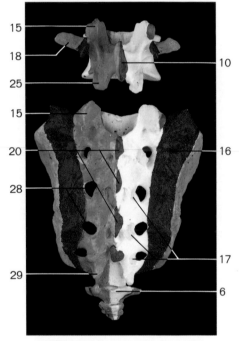

2.5   Articulation between two thoracic vertebrae (left lateral view).

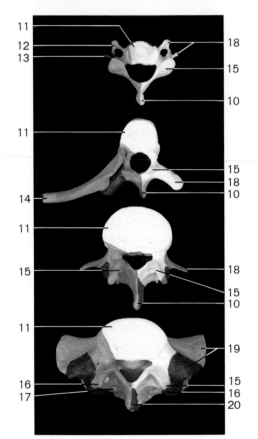

2.6   General characteristics of vertebrae. From top to bottom : Typical cervical, thoracic, lumbar and sacral vertebrae (superior view).

2.7   General characteristics of lumbar vertebra and sacrum (posterior view).

Blue: Mamillary processes
Green: Ribs or homologous processes
Red: Muscular processes (transverse and spinous processes)
Orange: Laminae and articular processes
Yellow: Articular processes

The axis (second cervical vertebra) has a dens projecting upward, which articulates with the atlas (first cervical vertebra) to serve as a pivot for rotational movement. The spinous process of the seventh cervical vertebra is long and prominent. This vertebra, called the *vertebra prominens*, is often used in living anatomy for determining vertebral order.

Vertebral articulations are movable because of intervertebral disks between the vertebral bodies. Mobility is greatest between cervical vertebrae and most difficult between thoracic vertebrae.

1	Cervical vertebrae	12	Anterior tubercle of transverse process
2	Thoracic vertebrae	13	Posterior tubercle of transverse process
3	Lumbar vertebrae		
4	Promontory of sacrum	14	Rib articulated with thoracic vetebra
5	Sacrum		
6	Coccyx	15	Superior articular process
7	Transverse foramen	16	Lateral sacral crest
8	Dens of axis	17	Intermediate sacral crest
9	Posterior tubercle of atlas	18	Transverse process
10	Spinous process	19	Lateral part of sacrum
11	Body of vertebra		

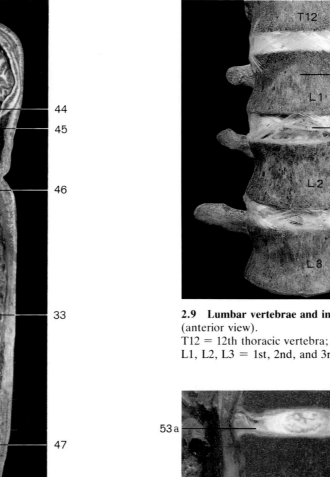

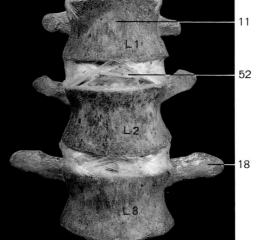

2.9 Lumbar vertebrae and intervertebral disks (anterior view).
T12 = 12th thoracic vertebra;
L1, L2, L3 = 1st, 2nd, and 3rd lumbar vertebrae.

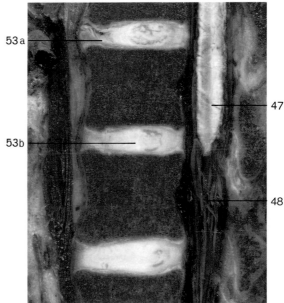

2.10 Longitudinal section through lumbar vertebrae and lowermost portion of the spinal cord.

2.8 Sagittal section through head and trunk (female). The conus medullaris of the spinal cord is located at the level of the first lumbar vertebra (L1).

20	Median sacral crest	26	Intervertebral foramen
21	Superior costal fovea (superior demifacet for rib head)	27	Costal fovea of transverse process
22	Inferior costal fovea (inferior demifacet for rib head)	28	Dorsal sacral foramina
		29	Sacral cornu
23	Intervertebral disk	30	Cerebrum
24	Superior and inferior vertebral notches	31	Larynx
		32	Trachea, thymus
25	Inferior articular process	33	Esophagus
		34	Heart
		35	Liver

36	Transverse colon	47	Conus medullaris
37	Stomach	48	Cauda equina
38	Pancreas	49	Rectum
39	Umbilicus	50	Vagina
40	Small intestine	51	Anus
41	Uterus	52	Fibrous ring of intervertebral disk
42	Urinary bladder		
43	Pubic symphysis	53	Intervertebral disk
44	Cerebrum		a Outer portion (Anulus fibrosus)
45	Medulla oblongata		
46	Spinal cord		b Inner core (Nucleus pulposus)

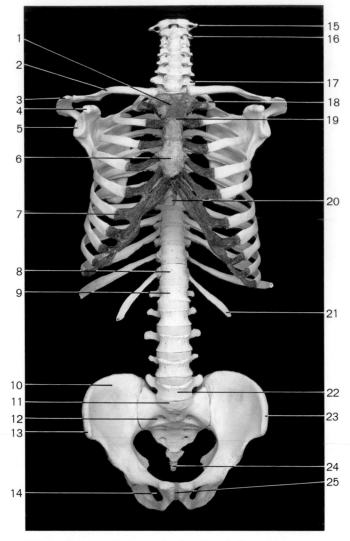

2.11A   Skeleton of the trunk, shoulder girdle, thorax, vertebral column and pelvis (anterior view).

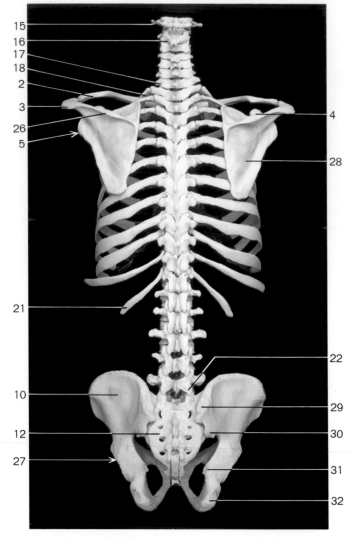

2.11B   Skeleton of the trunk, thorax, vertebral column and pelvis (posterior view).

The *true ribs,* first through seventh, articulate directly with the sternum via costal cartilages. The ribs from eighth to tenth are called *false ribs;* their cartilages do not articulate with the sternum but with the cartilages of the seventh and eighth ribs. The 11th and 12th ribs which are free at their anterior ends are called *floating ribs.*

The *sternal angle* (between manubrium and body), because it projects slightly anteriorly, can be palpated through the skin. It is at the level of the second costal cartilage and rib and provides a useful landmark in living anatomy for determining the order of ribs. The first rib is almost impossible to palpate.

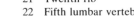

1	Manubrium of sternum	15	Atlas	29	Posterior superior iliac spine
2	Clavicle	16	Axis	30	Posterior inferior iliac spine
3	Acromion of scapula	17	Seventh cervical vertebra	31	Ischial spine
4	Coracoid process of scapula	18	First rib	32	Ischial tuberosity
5	Glenoid cavity	19	Sternal angle	33	Acromioclavicular joint
6	Body of sternum	20	Xiphoid process	34	Sternoclavicular joint
7	Costal cartilage	21	Twelfth rib	35	Sternum
8	Twelfth thoracic vertebra	22	Fifth lumbar vertebra	36	Supraglenoid tubercle
9	First lumbar vertebra	23	Iliac crest	37	Infraglenoid tubercle
10	Ilium	24	Coccyx	38	Lateral margin of scapula
11	Promontory of sacrum	25	Pubic symphysis	39	Costal arch
12	Sacrum	26	Spine of scapula	40	Supraspinous fossa of scapula
13	Anterior superior iliac spine	27	Acetabulum	41	Inferior angle of scapula
14	Obturator foramen	28	Infraspinous fossa of scapula	42	Angle of rib

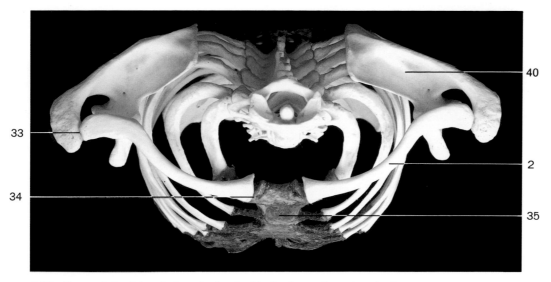

2.12 **Bones of shoulder girdle** articulated with the thorax (superior view).

The head of the rib articulates with the thoracic vertebra at two sites to form a gliding joint which enables the costal movement associated with breathing movements. The volume of the thoracic cavity increases when the ribs are raised due to contraction of the external intercostal muscles on inspiration. Conversely, the volume of the thoracic cavity decreases when the ribs are lowered on expiration.

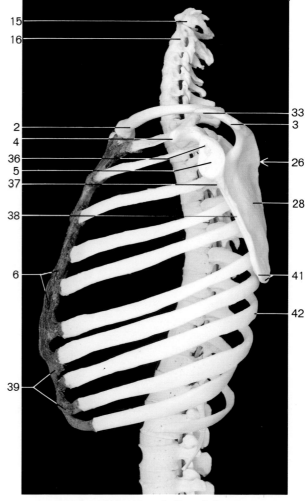

2.13 **Skeleton of shoulder girdle and thorax** (left lateral view).

3. Thorax, Abdomen and Back

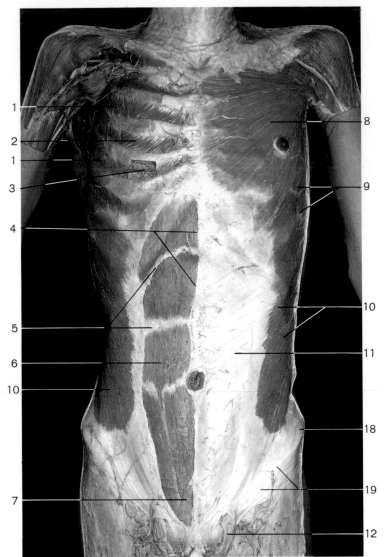

3.1 Muscles of the thoracic and abdominal wall (male). Sheath of the right rectus abdominis is opened.

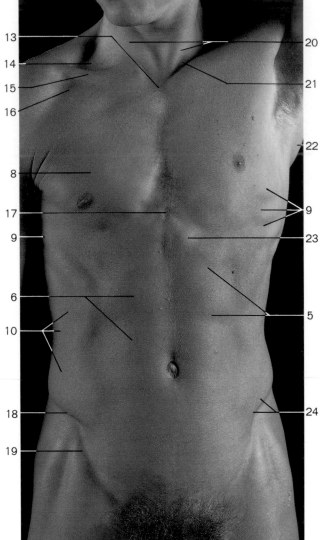

3.2 Surface anatomy of the thoracic and abdominal wall (anterior view).

1	External intercostal muscle
2	Internal intercostal muscle
3	Intercostal nerve, artery and vein
4	Linea alba
5	Tendinous intersections of rectus abdominis
6	Rectus abdominis
7	Pyramidalis
8	Pectoralis major
9	Serratus anterior
10	External oblique muscle
11	Anterior lamina of rectus sheath
12	Spermatic cord
13	Jugular fossa
14	Greater supraclavicular fossa
15	Clavicle
16	Deltopectoral triangle
17	Xiphoid process
18	Anterior superior iliac spine
19	Inguinal ligament
20	Sternocleidomastoid
21	Lesser supraclavicular fossa
22	Latissimus dorsi
23	Costal arch
24	Iliac crest

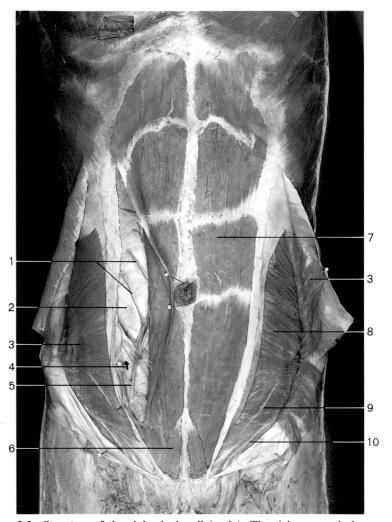

When the thoracic and abdominal wall muscles are compared, the external and internal intercostal muscles and transversus thoracis correspond to the external and internal oblique muscles and transversus abdominis, respectively.

1   Intercostal nerves, arteries and veins
2   Posterior lamina of rectus sheath
3   Internal oblique muscle
4   Arcuate line
5   Inferior epigastric artery and vein
6   Pyramidalis
7   Rectus abdominis
8   Transversus abdominis
9   Iliohypogastric nerve
10 Ilioinguinal nerve
11 Medial column of intrinsic muscles of the back
12 Lateral column of erector spinae
13 Thoracolumbar fascia with superficial and deep layer
14 External oblique muscle
15 Fascia transversalis
16 Anterior lamina of rectus sheath

3.3 Structure of the abdominal wall (male). The right external oblique is cut to show the internal oblique and the left internal oblique is removed to expose the transversus abdominis. The right rectus is reflected medially to expose the posterior lamina of the rectus sheath and arrangement of vessels and nerves of the rectus abdominis. The arcuate line (arrow) is the curved inferior terminal border of the posterior lamina of the rectus sheath.

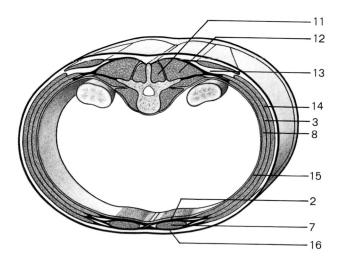

3.4 Cross-section of the trunk. Schematic drawing; superior to arcuate line.

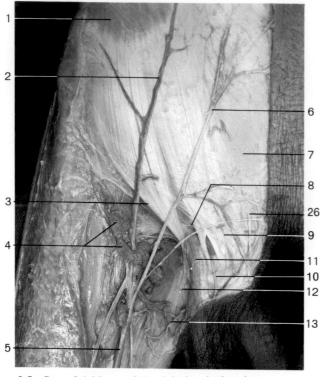

3.5   Superficial layer of the right inguinal region.

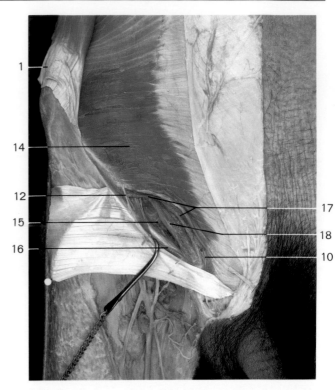

3.6   Deeper layer of the right inguinal region showing the inguinal canal.

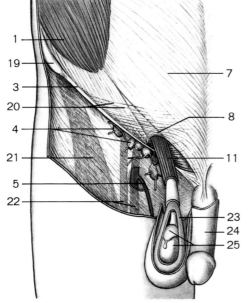

3.7   Lower part of anterior abdominal wall and inguinal canal. Schematic drawing.

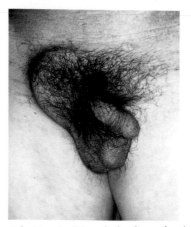

3.8   Inguinal hernia in the male. A part of the abdominal viscera protrudes through the inguinal canal into a space beneath the skin.

1 External oblique muscle	14 Internal oblique muscle
2 Superficial circumflex iliac artery and vein	15 Inferior epigastric artery and vein
3 Inguinal ligament	16 Cremaster muscle
4 Inguinal lymph nodes	17 Margin of transversus abdominis
5 Great saphenous vein	18 Ductus deferens
6 Superficial epigastric vein	19 Anterior superior iliac spine
7 Anterior lamina of rectus sheath	20 Intercrural fibers
8 Superficial inguinal ring	21 Fascia lata and sartorius muscle of thigh
9 Iliohypogastric nerve	22 Femoral artery and vein
10 Genitofemoral nerve	23 Tunica vaginalis
11 Spermatic cord	24 Penis
12 Ilioinguinal nerve	25 Epididymis and testis
13 External pudendal artery and vein	26 Subcostal nerve

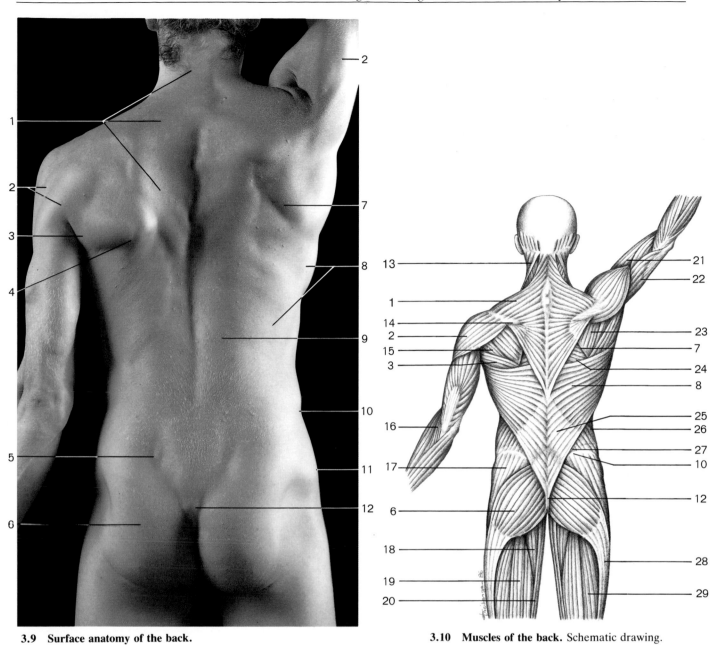

3.9 Surface anatomy of the back.

3.10 Muscles of the back. Schematic drawing.

1	Trapezius	11	Greater trochanter of femur
2	Deltoid	12	Coccyx
3	Teres major	13	Sternocleidomastoid
4	Inferior angle of scapula	14	Spine of scapula
5	Posterior superior iliac spine	15	Teres minor
6	Gluteus maximus	16	Anconeus
7	Medial margin of scapula	17	Gluteus medius
8	Latissimus dorsi	18	Adductor magnus
9	Erector spinae	19	Semimembranosus
10	Iliac crest	20	Gracilis

21	Biceps brachii
22	Triceps brachii
23	Infraspinatus
24	Rhomboideus major muscle
25	Thoracolumbar fascia
26	External oblique muscle
27	Lumbar triangle
28	Iliotibial tract
29	Long head of biceps femoris

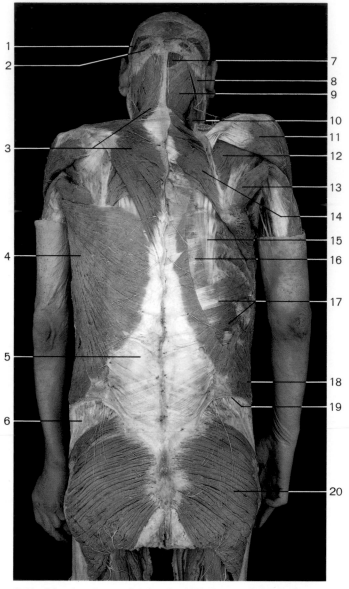

1 Occipital belly of occipitofrontalis muscle
2 Transverse muscle of nucha
3 Trapezius
4 Latissimus dorsi
5 Thoracolumbar fascia
6 Gluteus medius and fascia
7 Semispinalis capitis
8 Sternocleidomastoid
9 Splenius capitis
10 Levator scapulae
11 Deltoid
12 Infraspinatus
13 Teres major
14 Rhomboideus major
15 Iliocostalis thoracis
16 Longissimus thoracis
17 Inferior posterior serratus
18 External oblique muscle
19 Iliac crest
20 Gluteus maximus
21 Ventral root
22 Dorsal root and ganglion
23 Spinal cord
24 Medial and lateral cutaneous branches (rami) of spinal nerve
25 Posterior branch of spinal nerve
26 Anterior cutaneous branches of spinal nerve
27 Anterior branch of spinal nerve (intercostal nerve)
28 Lateral cutaneous branches of spinal nerve
29 Sympathetic ganglion and communicating rami
30 Sympathetic trunk
31 Spinal nerve

3.11   Muscles of superficial and middle layers of the back.

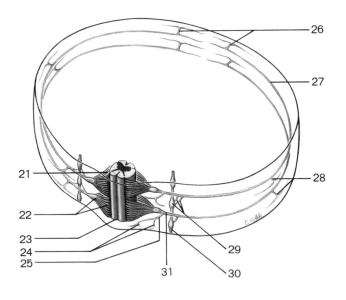

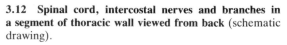

3.12   Spinal cord, intercostal nerves and branches in a segment of thoracic wall viewed from back (schematic drawing).

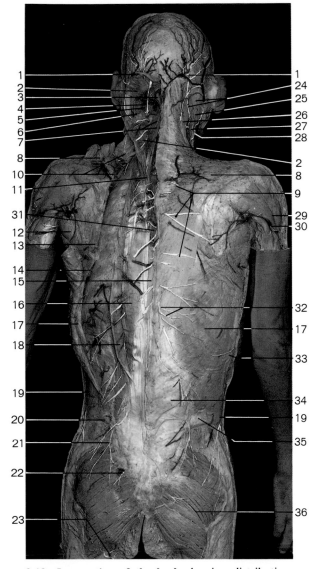

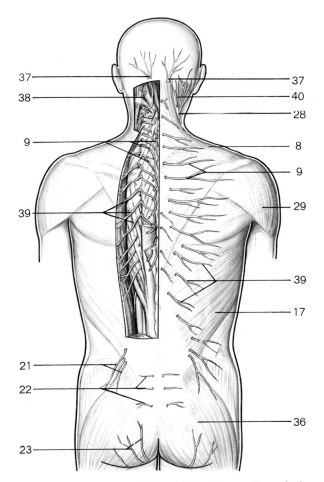

3.13 Innervation of the back showing distribution of the spinal nerves. On the left side, trapezius and latissimus dorsi have been removed. The semispinalis capitis is cut and reflected.

3.14 General characteristics of the innervation of the back, schematic drawing. Distribution of posterior branches of spinal nerves. Note the segmental arrangement of the innervation of the posterior part of the trunk.

1 Greater occipital nerve, occipital artery	21 Superior cluneal nerve (L1–L3)
2 Semispinalis capitis	22 Middle cluneal nerve (S1–S3)
3 Rectus capitis posterior major	23 Inferior cluneal nerve
4 Vertebral artery, suboccipital nerve	24 Lesser occipital nerve, sternocleidomastoid
5 Arch of atlas	25 Splenius capitis
6 Obliquus capitis superior	26 Third occipital nerve
7 Semispinalis cervicis	27 Posterior cutaneous branch of fourth cervical nerve
8 Trapezius	28 Great auricular nerve (C2, 3)
9 Medial cutaneous branches of dorsal rami of spinal nerves	29 Deltoid
10 Levator scapulae	30 Lateral cutaneous branch of axillary nerve
11 Longissimus cervicis	31 Multifidus
12 Teres major	32 Posterior cutaneous branch of twelfth thoracic nerve
13 Rhomboideus major	33 Lateral cutaneous branch of spinal nerve
14 Serratus anterior	34 Lumbar aponeurosis
15 Spinalis thoracis (upper half cut)	35 Lumbar triangle of Petit
16 Longissimus thoracis	36 Gluteus maximus
17 Latissimus dorsi	37 Greater occipital nerve (C2)
18 Iliocostalis	38 Suboccipital nerve (C1)
19 External oblique muscle	39 Lateral cutaneous branches of dorsal rami of spinal nerves
20 Iliac crest	40 Lesser occipital nerve (C2)

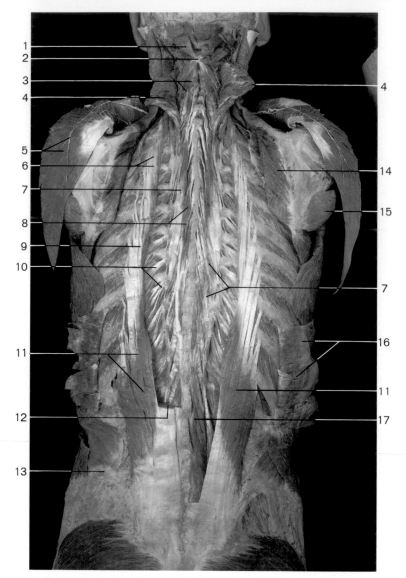

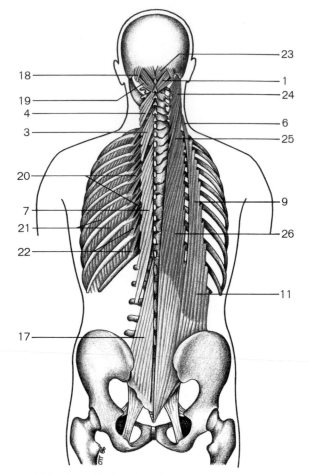

3.15   Deep muscles of the back. Longissimus dorsi are removed on both sides. Semispinalis of head has been cut and reflected.

3.16 Schematic drawing of the back muscles.

1	Rectus capitis posterior major	10	Levatores costarum	19	Obliquus capitis inferior
2	Spinous process of axis	11	**Iliocostalis lumborum**	20	Levatores costarum breves
3	Semispinalis cervicis	12	Longissimus thoracis (cut)	21	External intercostal muscles
4	Semispinalis capitis	13	Iliac crest	22	Levatores costarum longi
5	Trapezius (reflected), accessory nerve	14	Rhomboideus major (cut)	23	Rectus capitis posterior minor
6	**Iliocostalis cervicis**	15	Teres major	24	Longissimus capitis
7	Semispinalis thoracis	16	Inferior posterior serratus (reflected)	25	Longissimus cervicis
8	Spinalis thoracis	17	Multifidus	26	Longissimus thoracis
9	**Iliocostalis thoracis**	18	Obliquus capitis superior		

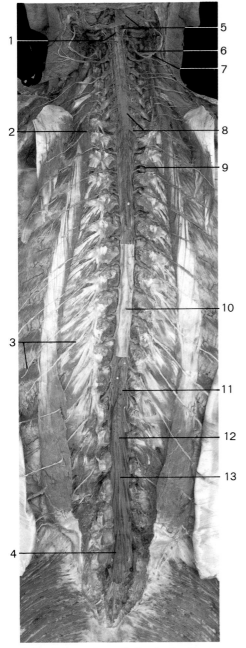

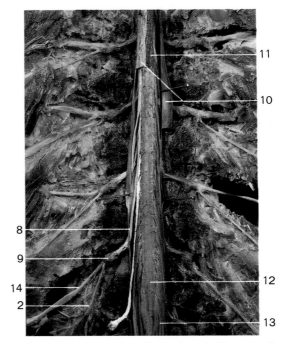

3.18 Lumbar portion of spinal cord. Note the discrepancy between the levels at which spinal nerve roots (indicated by different colors) arise from the spinal cord and the levels at which they exit via the intervertebral foramina.

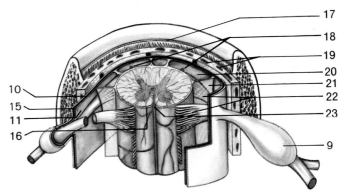

3.19 Meningeal coverings at the spinal cord, anterior view (schematic drawing).

3.17 Innervation of the back. Vertebral canal is opened to expose spinal cord. Longissimus dorsi is removed and iliocostalis reflected.

1 Medulla oblongata
2 Dorsal primary ramus
3 Lateral cutaneous branches of dorsal rami of spinal nerves
4 Filum terminale
5 Cerebellomedullary cistern and cerebellum
6 Greater occipital nerve
7 Third cervical nerve
8 Dorsal roots
9 Dorsal root ganglion
10 Dura mater
11 Spinal arachnoid
12 Conus medullaris
13 Cauda equina
14 Anterior branch of spinal nerve (intercostal nerve)
15 Subdural space
16 Pia mater
17 Periosteum of vertebral canal
18 Posterior spinal arteries
19 Epidural space with venous plexus and fatty tissue
20 Subarachnoid space
21 Denticulate ligament
22 Dorsal and ventral roots of spinal nerve
23 Anterior spinal artery

During early fetal development the spinal cord and vertebral column are almost the same length. Later as the vertebral column grows larger than the spinal cord, the spinal cord is pulled upward. In the upper segments spinal nerves emerge in an almost horizontal direction through intervertebral foramina, whereas in lower segments nerves gradually travel downward. At the end of the spinal cord the nerves extend vertically and surround the filum terminale to form the *cauda equina*. The lowermost part of the spinal cord in the adult extends to the level of the first or second lumbar vertebra and in the neonate it extends to the third lumbar vertebra.

4. Girdles and Free Extremities

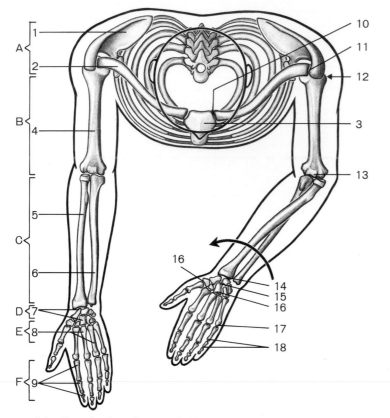

A   Pectoral girdle
B   Upper arm
C   Forearm
D   Wrist
E   Palm of hand
F   Fingers

 1   Scapula
 2   Clavicle
 3   Sternum
 4   Humerus
 5   Radius
 6   Ulna
 7   Carpal bones
 8   Metacarpal bones
 9   Phalanges
10   Sternoclavicular joint
11   Acromioclavicular joint
12   Shoulder joint
13   Elbow joint
14   Radiocarpal joint
15   Midcarpal joint
16   Carpometacarpal joint
17   Metacarpophalangeal joint
18   Interphalangeal joints of fingers
19   Acromion of scapula
20   Greater tubercle of humerus
21   Head of humerus
22   Glenoid cavity
23   Surgical neck of humerus
24   Scapular notch
25   Coracoid process of scapula
26   Infraglenoid tubercle
27   Anatomical neck of humerus
28   Spine of scapula

4.1   Organization of pectoral (shoulder) girdle and upper limb (superior and anterior views). Two positions of the forearm that are essential to manual skills in humans, supination (right arm) and pronation (left arm), are shown.

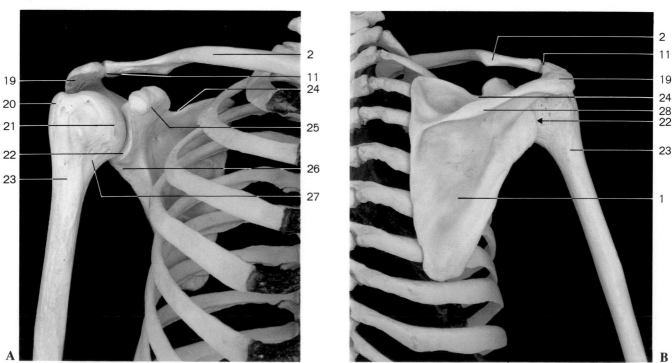

4.2   Bones of shoulder joint, anterior (A) and posterior (B) views.

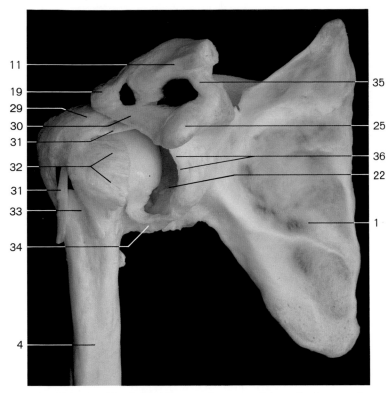

4.3 Right shoulder joint. The anterior part of the articular capsule has been removed and the head of the humerus has been slightly rotated outward to show the cavity of the joint.

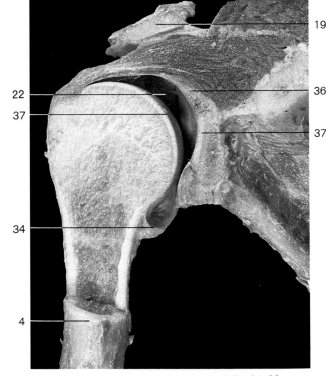

4.4 Frontal (coronal) section through right shoulder (anterior view).

The shoulder joint is a ball-and-socket type joint and allows for the greatest mobility among the synovial joints due to its shallow glenoid cavity.

Mobility and strength are increased by the **musculo-tendinous cuff**; known as the rotator cuff, which is formed by the muscle and inserting tendon fibers of the supraspinatus, infraspinatus, teres minor and sub-scapularis muscles, that blend with the articular capsule.

Further, movement of the shoulder girdle supplements mobility of the humerus resulting in its maximal movement.

29 Tendon of supraspinatus muscle (attached to articular capsule)
30 Coracoacromial ligament
31 Tendon of long head of biceps brachii
32 Tendon of subscapularis (attached to articular capsule)
33 Intertubercular groove of humerus
34 Articular capsule of shoulder joint
35 Trapezoid ligament
36 Glenoid labrum
37 Articular cartilage
38 Crest of greater tubercle
39 Deltoid tuberosity
40 Radial fossa
41 Lateral epicondyle of humerus
42 Capitulum of humerus
43 Lesser tubercle
44 Crest of lesser tubercle
45 Coronoid fossa
46 Medial epicondyle of humerus
47 Trochlea
48 Shaft of humerus
49 Ulnar nerve sulcus
50 Radial nerve sulcus
51 Olecranon fossa

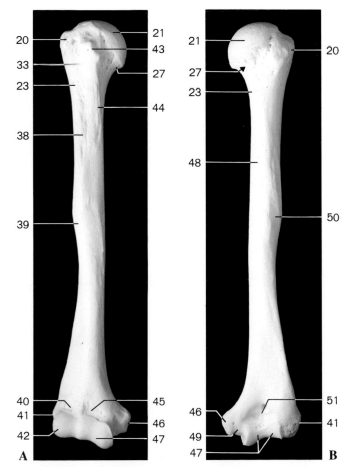

4.5 Right humerus, anterior (A) and posterior (B) views.

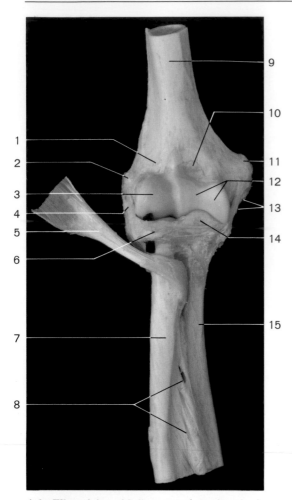

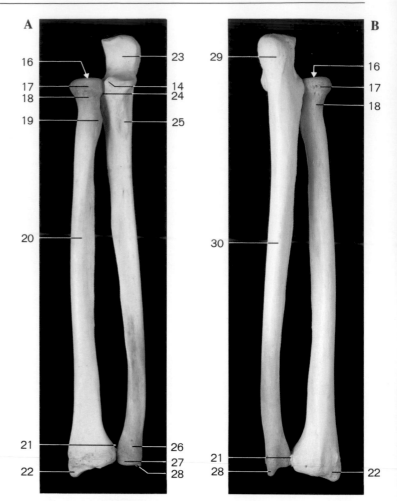

4.6   Elbow joint with ligaments (anterior view). Articular capsule has been removed to show the annular ligament.

4.7   Bones of right forearm, radius and ulna, anterior (A) and posterior (B) views.

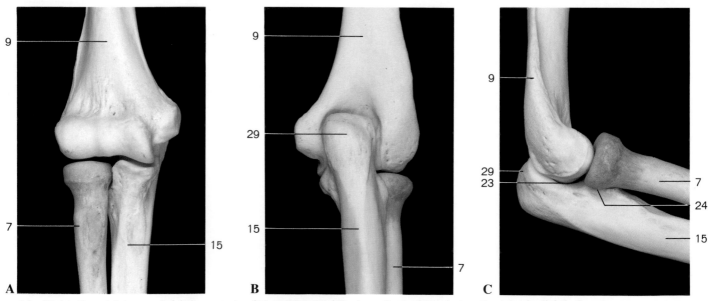

4.8   Right elbow joint, anterior (A), posterior (B) and lateral (C) views. In the lateral view the joint is slightly flexed.

The forearm bones articulate with the carpal bones forming ellipsoidal joints which limit movement of the hand to two directions, lateral and anteroposterior. The radius and ulna lie parallel to each other or the radius crosses over the ulna as shown, so rotation of the hand can be achieved. With the palm facing anteriorly in supination the forearm bones lie parallel; with the palm facing posteriorly in pronation the forearm bones are crossed when the pivot joints at both ends of the forearm bones work.

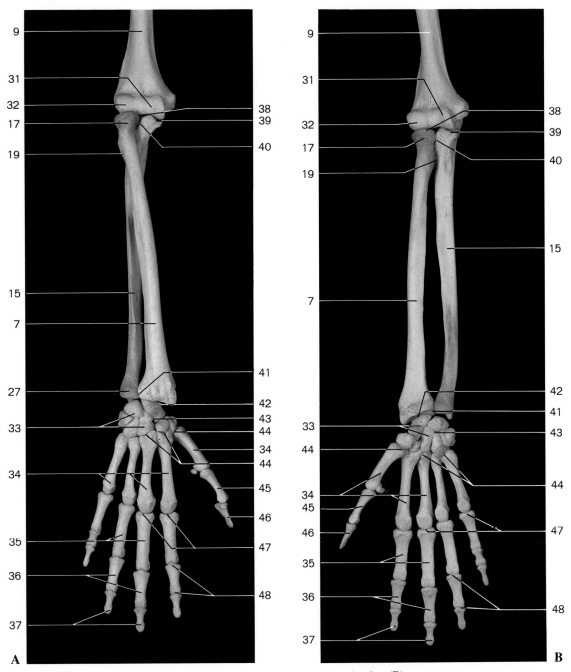

4.9 **Skeleton of right forearm and hand in pronation (A) and supination (B).**

1	Radial fossa	12	Trochlea of humerus
2	Lateral epicondyle of humerus	13	Ulnar collateral ligament
3	Capitulum of humerus	14	Coronoid process of ulna
4	Radial collateral ligament	15	Ulna
5	Tendon of biceps brachii	16	Head of radius
6	Annular ligament of radius	17	Articular circumference of radius
7	Radius	18	Neck of radius
8	Interosseous membrane	19	Radial tuberosity
9	Humerus	20	Shaft of radius
10	Coronoid fossa	21	Ulnar notch of radius
11	Medial epicondyle of humerus	22	Styloid process of radius
		23	Trochlear notch
		24	Radial notch of ulna
		25	Ulnar tuberosity

26	Head of ulna	40	Proximal radioulnar joint
27	Articular. circumference of ulna	41	Distal radioulnar joint
28	Styloid process of ulna	42	Radiocarpal joint
29	Olecranon process	43	Midcarpal joint
30	Shaft of ulna	44	Carpometacarpal joint
31	Trochlca	45	Proximal phalanx of thumb
32	Capitulum	46	Distal phalanx of thumb
33	Carpal bones	47	Metacarpophalangeal joints
34	Metacarpal bones	48	Interphalangeal joints of fingers
35	Proximal phalanges		
36	Middle phalanges		
37	Distal phalanges		
38	Humeroradial joint		
39	Humeroulnar joint		

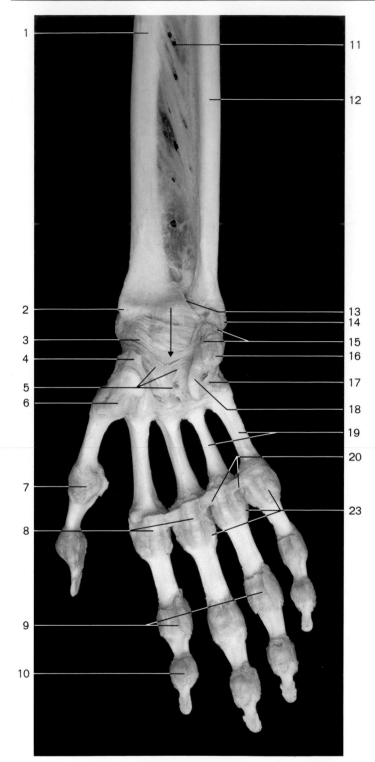

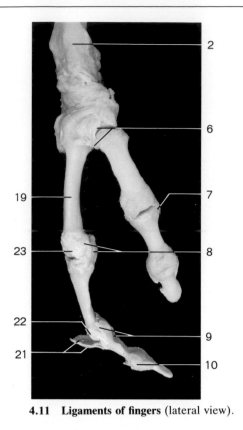

4.11 Ligaments of fingers (lateral view).

1   Radius
2   Styloid process of radius
3   Palmar radiocarpal ligament
4   Tendon of flexor carpi radialis (cut)
5   Radiating carpal ligament
6   Articular capsule of carpometacarpal joint of thumb
7   Articular capsule of metacarpophalangeal joint of thumb
8   Palmar ligaments and articular capsule of metacarpophalangeal joints
9   Palmar ligaments and articular capsule of interphalangeal joints
10 Articular capsules
11 Interosseous membrane
12 Ulna
13 Distal radioulnar joint
14 Styloid process of ulna
15 Palmar ulnocarpal ligament
16 Pisiform bone with tendon of flexor carpi ulnaris
17 Pisometacarpal ligament
18 Pisohamate ligament
19 Metacarpal bones
20 Deep transverse metacarpal ligament
21 Tendons of extensor muscles and articular capsule (cut)
22 Collateral ligament of interphalangeal joint
23 Collateral ligaments of metacarpophalangeal joints

4.10 Ligaments of right forearm, hand and fingers (palmar view). The arrow indicates location of the carpal tunnel.

The second through fifth carpometacarpal articulations, which are gliding joints, have limited mobility. The carpometacarpal articulation of the thumb (between trapezium and first metacarpal bone), which is a saddle joint, has greater mobility. Because the direction of this carpometacarpal joint is not parallel with the palmar plane it facilitates gripping (opposition).

The movement of intercarpal articulations is minimal due to several ligaments; however, the sliding carpal bones enable adduction and abduction of the hand.

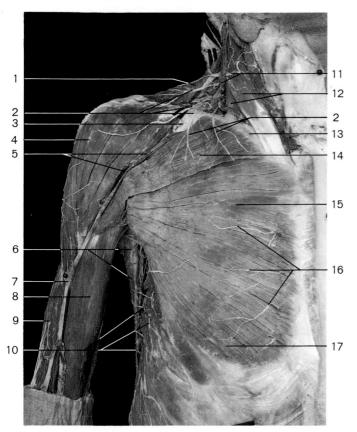

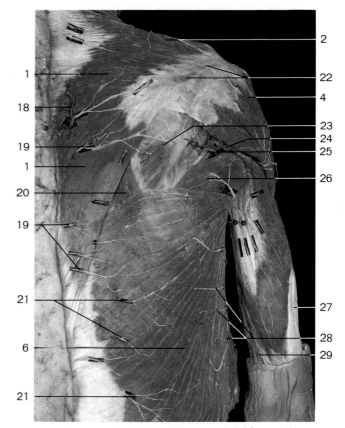

4.12 **Right shoulder and thoracic wall**, superficial layer (anterior view). Dissection of the cutaneous nerves and veins.

4.13 **Dorsal region of shoulder**, superficial layer. Note segmental arrangement of the cutaneous nerves of the back.

1 Trapezius
2 Supraclavicular nerves
3 Deltopectoral triangle
4 Deltoid muscle
5 Cephalic vein within deltopectoral groove
6 Latissimus dorsi
7 Cephalic vein within lateral bicipital groove
8 Biceps brachii
9 Triceps brachii
10 Lateral cutaneous branches of intercostal nerves
11 Transverse cervical nerve and external jugular vein
12 Sternocleidomastoid
13 Clavicle
14 Clavicular part of pectoralis major
15 Sternocostal part of pectoralis major
16 Anterior cutaneous branches of intercostal nerves
17 Abdominal part of pectoralis major
18 Posterior branches of posterior intercostal artery and vein (medial cutaneous branches)
19 Medial cutaneous branches of dorsal rami of spinal nerves
20 Rhomboideus major
21 Lateral cutaneous branches of dorsal rami of spinal nerves
22 Spine of scapula
23 Infraspinatus and fascia
24 Triangular space with circumflex scapular artery and vein
25 Teres minor
26 Teres major
27 Tendon of triceps brachii
28 Lateral cutaneous branches of intercostal nerves
29 Medial cutaneous nerve of forearm
30 Posterior fibers of deltoid muscle
31 Long head of triceps brachii
32 Lateral head of triceps brachii
33 Ulnar nerve

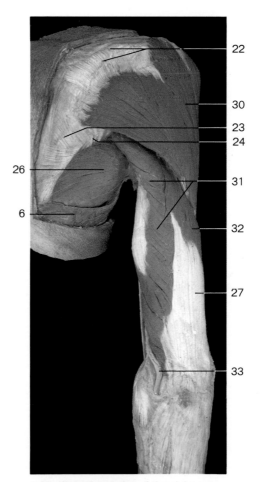

4.14 **Dorsal muscles of the right arm**, superficial layer (posterior view).

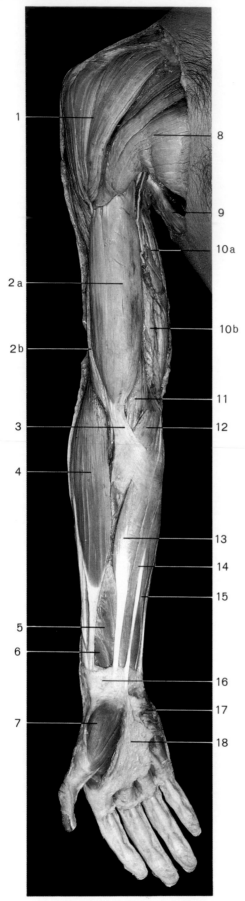

4.15 Superficial muscles of right upper limb (anterior view).

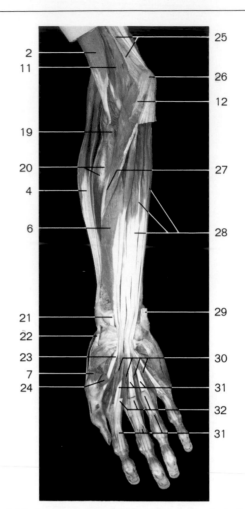

4.16 Forearm muscles, middle layer (anterior view). The palmaris longus, flexor carpi radialis and ulnaris have been partly removed. The flexor retinaculum has been divided.

1 Deltoid
2 Biceps brachii
    a Short head
    b Long head
3 Bicipital aponeurosis
4 Brachioradialis
5 Superficial flexor muscle of fingers
6 Flexor pollicis longus
7 Abductor pollicis brevis
8 Pectoralis major
9 Latissimus dorsi
10 Triceps brachii
    a Long head
    b Medial head
11 Brachialis
12 Pronator teres muscle (humeral head)
13 Flexor carpi radialis
14 Palmaris longus
15 Flexor carpi ulnaris
16 Flexor retinaculum
17 Palmaris brevis
18 Palmar aponeurosis
19 Supinator

20 Radius and extensor carpi radialis brevis
21 Tendon of flexor carpi radialis (cut)
22 Tendon of abductor pollicis longus
23 Tendon of flexor pollicis longus
24 Superficial head of flexor pollicis brevis (cut)
25 Medial intermuscular septum
26 Medial epicondyle of humerus
27 Pronator teres (insertion on radius)
28 Flexor digitorum profundus
29 Tendon of flexor carpi ulnaris (cut)
30 Lumbrical muscles
31 Tendons of flexor digitorum profundus
32 Tendons of flexor digitorum superficialis (cut)

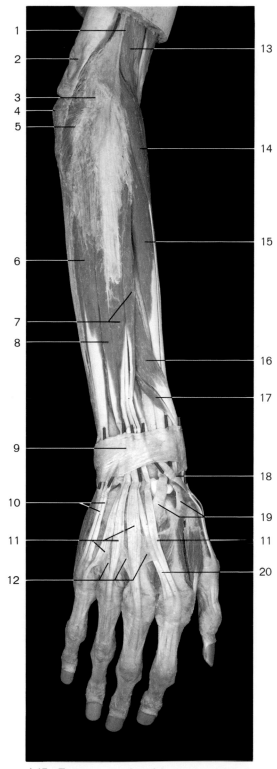

4.17 Extensor muscles of forearm and hand, superficial layer (posterior view). Tunnels for extensor tendons indicated by probes.

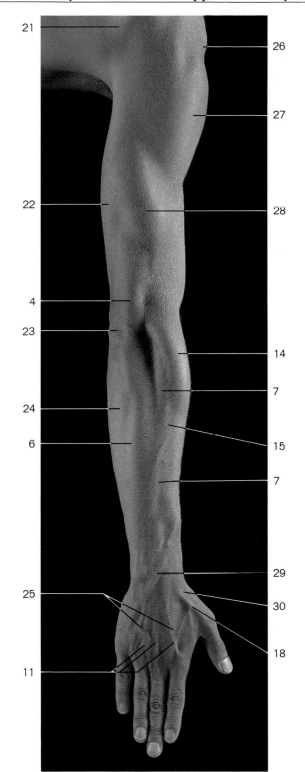

4.18 Surface anatomy of right upper extremity (posterior view).

1	Lateral intermuscular septum	10	Tendons of extensor digiti minimi
2	Tendon of triceps brachii	11	Tendons of extensor digitorum
3	Lateral epicondyle of humerus	12	Intertendinous connections
4	Olecranon process of ulna	13	Brachioradialis
5	Anconeus	14	Extensor carpi radialis longus
6	Extensor carpi ulnaris	15	Extensor carpi radialis brevis
7	Extensor digitorum	16	Abductor pollicis longus
8	Extensor digiti minimi		
9	Extensor retinaculum		

17	Extensor pollicis brevis	24	Flexor carpi ulnaris
18	Tendon of extensor pollicis longus	25	Venous network
19	Tendons of extensor carpi radialis longus and brevis	26	Acromion process
20	Tendon of extensor indicis	27	Deltoid
21	Teres major	28	Lateral head of triceps brachii
22	Medial head of triceps brachii	29	Cephalic vein
23	Site for palpation of ulnar nerve	30	Anatomical snuff box

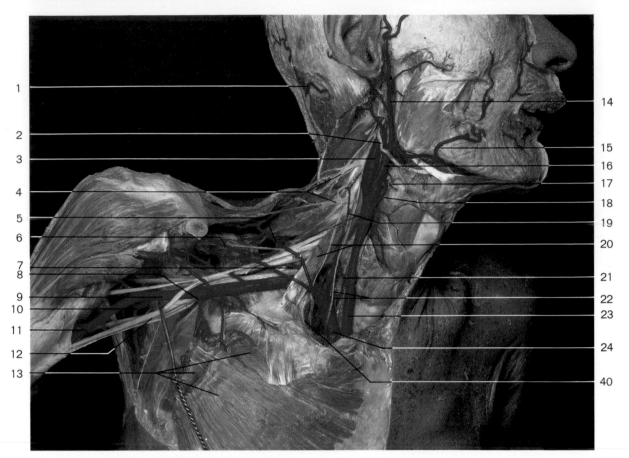

4.19   Main branches of right subclavian and axillary arteries (anterior view). The clavicle has been cut.

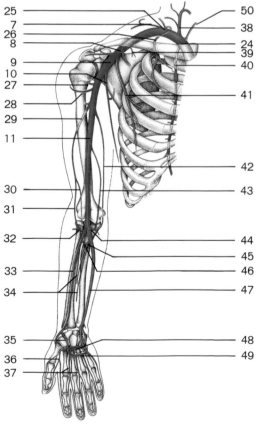

4.20   Arteries of right arm, anterior view
(semischematic drawing).

1 Occipital artery	26 Highest intercostal artery
2 Hypoglossal nerve	27 Posterior circumflex humeral
3 Internal carotid artery	artery
4 Cervical plexus	28 Anterior circumflex humeral
5 Superficial cervical artery	artery
6 Suprascapular artery	29 Profunda brachii artery
7 Deep branch of transverse	30 Middle collateral artery
cervical artery (descending	31 Radial collateral artery
scapular artery)	32 Radial recurrent artery
8 Thoracoacromial artery	33 **Radial artery**
9 **Axillary artery**	34 Anterior and posterior inter-
10 Subscapular artery	osseous arteries
11 **Brachial artery**	35 Superficial palmar branch of
12 Brachial plexus	radial artery
13 Pectoralis major and minor	36 Princeps pollicis artery
14 Superficial temporal artery	37 Common palmar digital arteries
15 Facial artery	38 Thyrocervical trunk
16 External carotid artery	39 Highest thoracic artery
17 Superior laryngeal artery	40 Internal thoracic artery
18 Superior thyroid artery	41 Lateral thoracic artery
19 Ascending cervical artery	42 Superior ulnar collateral artery
20 Scalenus anterior muscle,	43 Inferior ulnar collateral artery
phrenic nerve	44 Ulnar recurrent artery
21 Inferior thyroid artery	45 Common interosseous artery
22 Vertebral artery, vagus nerve	46 Recurrent interosseous artery
23 Right common carotid artery	47 **Ulnar artery**
24 **Right subclavian artery**	48 Deep palmar arch
25 Costocervical trunk	49 Superficial palmar arch
	50 Vertebral artery

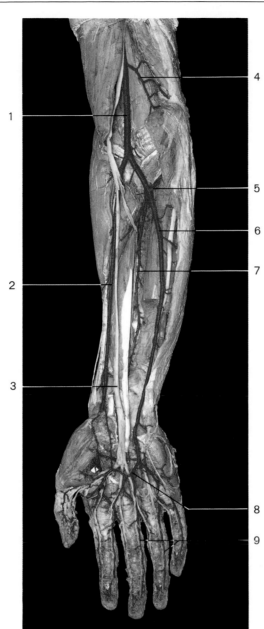

4.21 Arteries and nerves of right forearm and hand (anterior view).

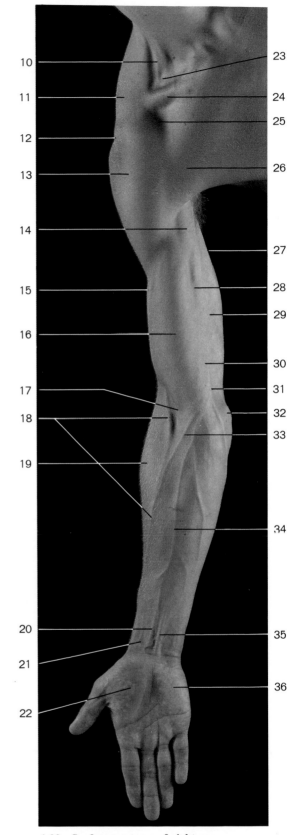

4.22 Surface anatomy of right arm (anterior view).

1 Brachial artery
2 Radial artery
3 Median nerve
4 Inferior ulnar collateral artery
5 Ulnar recurrent artery
6 Ulnar artery
7 Anterior interosseous artery
8 Superficial palmar arch
9 Proper palmar digital arteries
10 Platysma
11 Trapezius
12 Acromion process
13 Deltoid
14 Coracobrachialis
15 Lateral head of triceps brachii
16 Biceps brachii
17 Cubital fossa
18 Cephalic vein
19 Brachioradialis
20 Tendon of flexor carpi radialis
21 Radial antebrachial sulcus
22 Thenar
23 Supraclavicular fossa
24 Clavicle
25 Deltopectoral triangle
26 Pectoralis major
27 Long head of triceps brachii
28 Medial bicipital sulcus
29 Medial head of triceps brachii
30 Brachialis
31 Basilic vein
32 Medial epicondyle of humerus
33 Median cubital vein
34 Median vein of forearm
35 Tendon of palmaris longus
36 Hypothenar

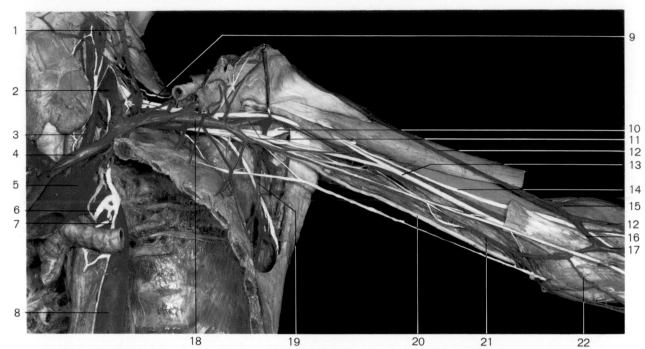

4.23 **Veins and nerves of left arm** (anterior view). The clavicle and the anterior wall of the thorax have been cut. The shoulder is slightly reflected.   Green = thoracic duct;   white = nerves;   yellow = sympathetic cardiac nerve.

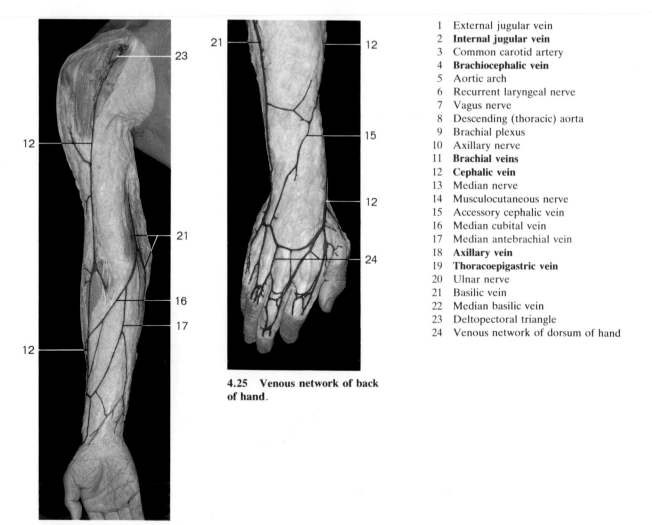

4.24 **Superficial veins of right arm**.

4.25 **Venous network of back of hand**.

1   External jugular vein
2   **Internal jugular vein**
3   Common carotid artery
4   **Brachiocephalic vein**
5   Aortic arch
6   Recurrent laryngeal nerve
7   Vagus nerve
8   Descending (thoracic) aorta
9   Brachial plexus
10 Axillary nerve
11 **Brachial veins**
12 **Cephalic vein**
13 Median nerve
14 Musculocutaneous nerve
15 Accessory cephalic vein
16 Median cubital vein
17 Median antebrachial vein
18 **Axillary vein**
19 **Thoracoepigastric vein**
20 Ulnar nerve
21 Basilic vein
22 Median basilic vein
23 Deltopectoral triangle
24 Venous network of dorsum of hand

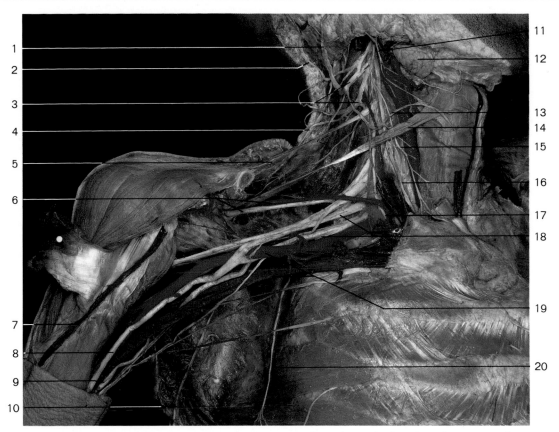

4.26   Right axillary region (anterior view). The pectoralis major and minor have been cut and reflected to expose the axillary nerves and vessels.

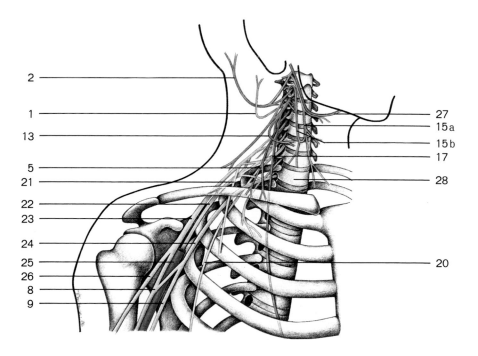

4.27   Main branches of cervical and brachial plexuses (schematic drawing). Brown = cervical plexus;   yellow = brachial plexus;   green = hypoglossal nerve.

1   Great auricular nerve
2   Lesser occipital nerve
3   **Cervical plexus**
4   Accessory nerve
5   Supraclavicular nerves
6   Suprascapular artery and nerve
7   Cephalic vein (cut and reflected)
8   Median nerve
9   Ulnar nerve
10   Thoracodorsal nerve
11   Cervical branch of facial nerve
12   Submandibular gland
13   Transverse cervical nerve
14   Omohyoid muscle
15   Ansa cervicalis
     a   Superior root
     b   Inferior root
16   Vagus nerve
17   Phrenic nerve
18   **Brachial plexus**
19   Axillary artery and vein
20   Long thoracic nerve
21   Lateral pectoral nerve
22   Posterior cord of brachial plexus
23   Lateral cord of brachial plexus
24   Medial cord of brachial plexus
25   Axillary artery
26   Musculocutaneous nerve
27   Hypoglossal nerve
28   First thoracic vertebra

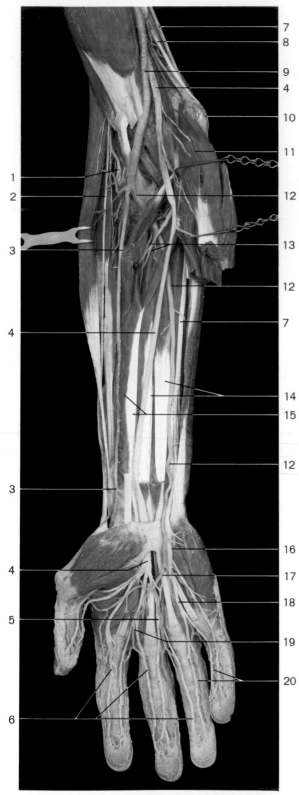

4.28   Vessels and nerves of forearm and hand, deep layer (anterior view). The superficial layer of flexor muscles has been removed.

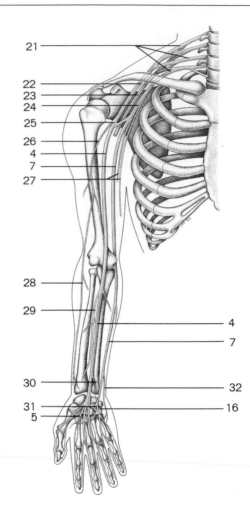

4.29   Main branches of musculo-cutaneous, median and ulnar nerves (schematic drawing).

1 Deep branch of radial nerve	19 Common palmar digital arteries
2 Superficial branch of radial nerve	20 Palmar digital branches of ulnar nerve
3 Radial artery	21 Brachial plexus
4 **Median nerve**	22 Lateral cord of brachial plexus
5 Common palmar digital branch of median nerve	23 Posterior cord of brachial plexus
6 Palmar digital nerves (median nerve)	24 Medial cord of brachial plexus
7 **Ulnar nerve**	25 Roots of median nerve
8 Medial intermuscular septum of arm	26 **Musculocutaneous nerve**
9 Brachial artery	27 Medial cutaneous nerves of arm and forearm
10 Medial epicondyle of humerus	28 Lateral cutaneous nerve of forearm
11 Pronator teres	29 Anterior interosseous nerve
12 Ulnar artery	
13 Anterior interosseous artery and nerve	30 Palmar branch of median nerve
14 Flexor digitorum profundus	31 Deep branch of ulnar nerve
15 Flexor pollicis longus	
16 Superficial branch of ulnar nerve	32 Dorsal branch of ulnar nerve
17 **Superficial palmar arch**	
18 Common palmar digital branch of ulnar nerve	

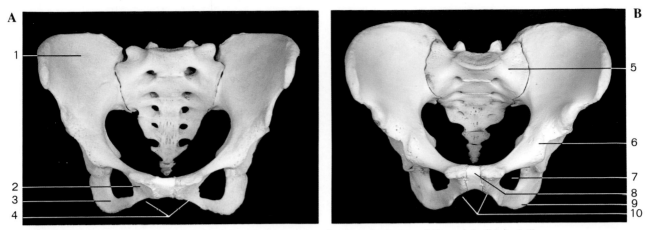

4.30 **Anterior view of the female (A) and male (B) pelvises** showing the bones of the pelvic (hip) girdles.

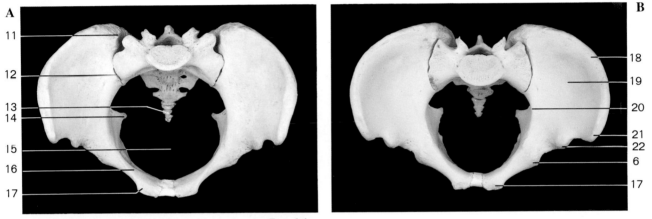

4.31 **Superior view of the female (A) and male (B) pelvises**.

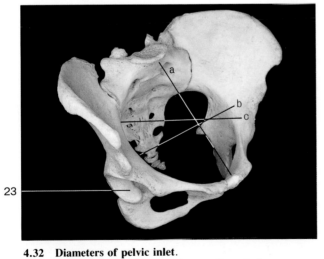

4.32 **Diameters of pelvic inlet**.
  a Anteroposterior (true conjugate diameter)
  b Transverse
  c Oblique

1   Ilium
2   Pubis
3   Ischium
4   Pubic arch
5   Sacrum
6   Iliopubic eminence
7   Obturator foramen
8   Pubic symphysis
9   Ischial tuberosity
10 Subpubic angle
11 Posterior superior iliac spine
12 Sacroiliac joint
13 Coccyx
14 Spine of ischium
15 Pelvic cavity
16 Pecten
17 Pubic tubercle
18 Iliac crest
19 Iliac fossa
20 Arcuate line of ilium
21 Anterior superior iliac spine
22 Anterior inferior iliac spine
23 Acetabulum

The female pelvis is broader and more flattened than the male pelvis for adaptation of the pelvic cavity to allow for passage of the fetal head in childbirth. The pelvic cavity is larger in the female. The superior aperture of the pelvic inlet is ellipsoidal in the female and heart-shaped in the male. The right and left inferior branches of the pubic bones form a wider angle in the female (pubic arch) and a more acute angle in the male (subpubic angle).

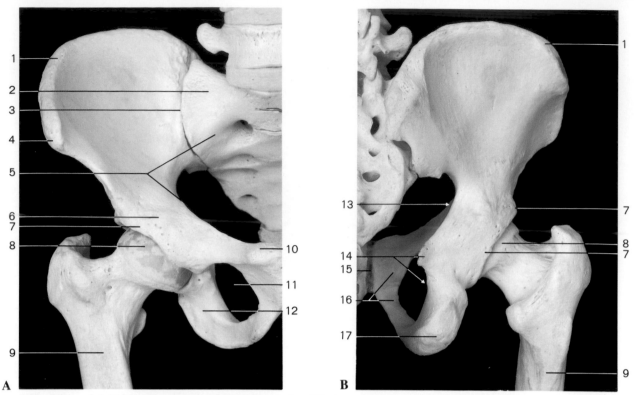

4.33   Bones of right hip joint, anterior (A) and posterior (B) views.

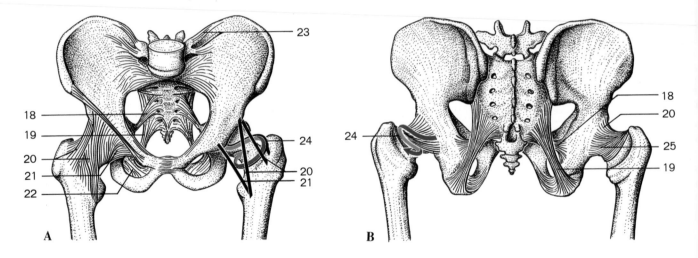

4.34   Ligaments of pelvis and hip joint, anterior (A) and posterior (B) views.

1 Iliac crest	10 Pubic tubercle	18 Sacrospinal ligament
2 Lateral part of sacrum	11 Obturator foramen	19 Sacrotuberal ligament
3 Sacroiliac joint	12 Ramus of ischium	20 Iliofemoral ligament
4 Superior anterior iliac spine	13 Greater sciatic notch of ilium	21 Pubofemoral ligament
5 Linea terminalis	14 Spine of ischium, lesser sciatic notch	22 Obturator membrane
6 Iliopubic eminence	15 Pubic symphysis	23 Iliolumbar ligament
7 Bony margin of acetabulum	16 Pubis	24 Orbicular zone
8 Head of femur	17 Ischial tuberosity	25 Ischiofemoral ligament
9 Femur		

Like the shoulder joint the hip joint is also a ball-and-socket type joint; however, its mobility is more limited because of the depth of the acetabulum which helps support the body weight. Congenital dislocation of the hip is caused by undergrowth of the upper portion of the socket. The ligament of the femoral head carries blood vessels which partially supply the head of the femur.

The greater trochanter which can be palpated through the skin is used, as is the anterior superior iliac spine, as a landmark when measuring the length of the lower extremity.

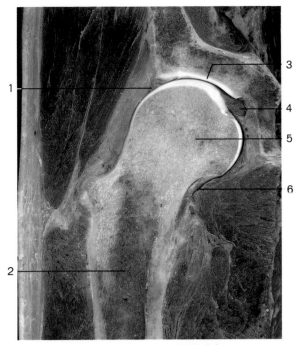

4.35 **Frontal (coronal) section of hip joint**.

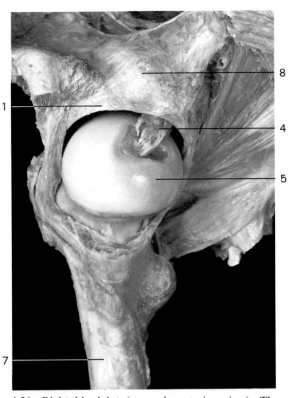

4.36 **Right hip joint** (opened, anterior view). The femur has been supinated. The ligament of the femoral head does not have mechanical functions; it carries blood vessels (branches of obturator artery) that nourish the head.

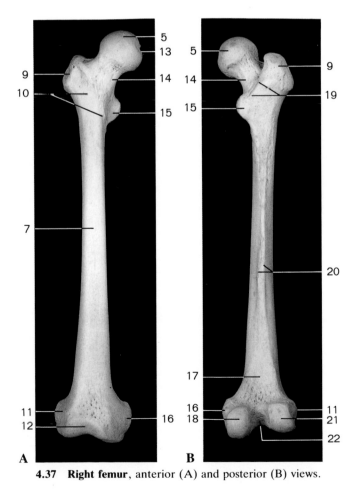

4.37 **Right femur**, anterior (A) and posterior (B) views.

1	Acetabular lip
2	Femur
3	Acetabulum
4	Ligament of head of femur
5	Head of femur
6	Articular cavity of hip joint
7	Shaft of femur
8	Body of ilium
9	Greater trochanter
10	Intertrochanteric line
11	Lateral epicondyle
12	Patellar surface
13	Fovea of head
14	Neck of femur
15	Lesser trochanter
16	Medial epicondyle
17	Popliteal surface
18	Medial condyle
19	Intertrochanteric crest
20	Linea aspera
21	Lateral condyle
22	Intercondylar fossa

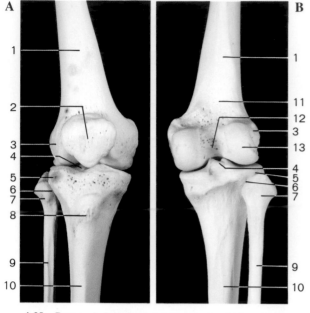

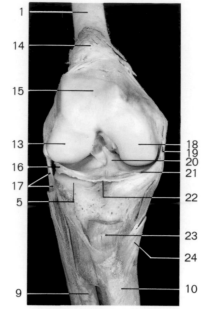

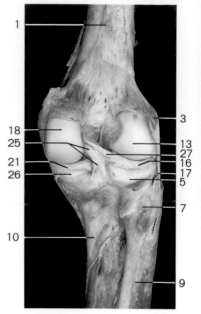

4.38   Bones of right knee joint, anterior (A) and posterior (B) views.

4.39A   Right knee joint (opened) **with ligaments** (anterior view). The patella and articular capsule have been removed and the femur slightly flexed.

4.39B Right knee joint with ligaments (posterior view). The joint is extended and the articular capsule has been removed.

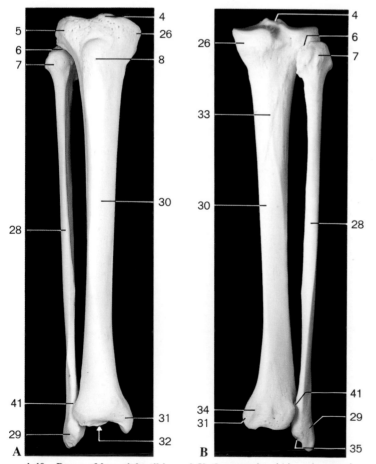

4.40   Bones of leg, right tibia and fibula, anterior (A) and posterior (B) views.

1   **Femur**
2   Patella
3   Lateral epicondyle of femur
4   Intercondylar eminence of tibia
5   Lateral condyle of tibia
6   Proximal tibiofibular joint
7   Head of fibula
8   Tibial tuberosity
9   **Fibula**
10   **Tibia**
11   Popliteal surface of femur
12   Intercondylar fossa of femur
13   Lateral condyle of femur
14   Articular capsule with suprapatellar bursa
15   Patellar surface
16   Lateral meniscus of knee joint
17   Fibular collateral ligament
18   Medial condyle of femur
19   Tibial collateral ligament
20   Anterior cruciate ligament
21   Medial meniscus of knee joint
22   Transverse ligament of knee
23   Patellar ligament
24   Common tendon of sartorius, semitendinosus and gracilis
25   Posterior cruciate ligament
26   Medial condyle of tibia
27   Posterior meniscofemoral ligament
28   Shaft of fibula
29   Lateral malleolus of fibula
30   Shaft of tibia
31   Medial malleolus of tibia
32   Inferior articular surface of tibia
33   Soleal line of tibia
34   Malleolar sulcus of tibia
35   Malleolar articular surface of fibula

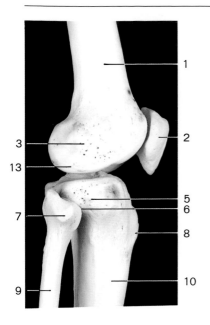

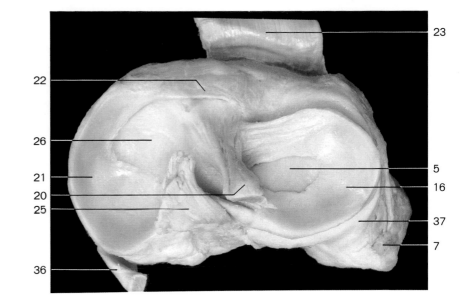

4.41 Bone of right knee joint (lateral view).

4.42 Articular surface of right tibial menisci and cruciate ligaments (superior view). Anterior margin of tibia on top.

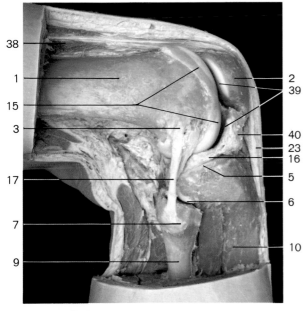

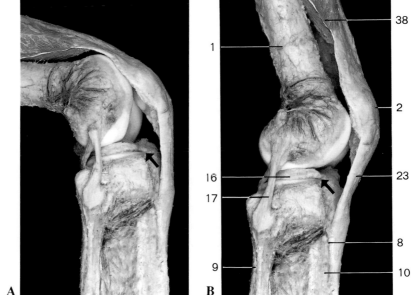

4.43 Right knee joint and tibiofibular joint with ligaments. Note location of the lateral meniscus.

4.44 Right knee joint (lateral view) with the leg flexed (A) and extended (B). With the flexion, the lateral meniscus slips backward (arrows).

36 Tendon of semimembranosus muscle
37 Posterior attachment of articular capsule of knee joint
38 Quadriceps femoris
39 Articular cavity of knee joint
40 Infrapatellar fat pad
41 Distal tibiofibular joint

In the knee joint two fibrocartilaginous rings (menisci) are attached to the superior articular surface of the tibia. Here they form a slightly elevated rim for better articulation with the lower end of the femur. This is due in part to the irregular concavity of the inferior articular surface of the femur. A pair of cruciate ligaments that run longitudinally and obliquely in the knee joint prevents the tibia from slipping forward or backward, and is under maximal tension when the joint is extended.

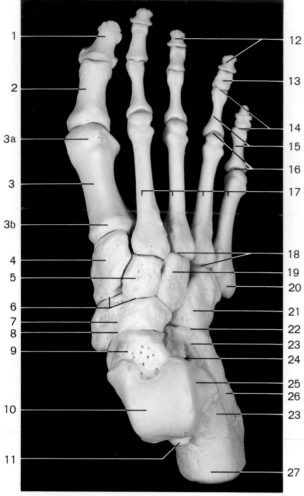

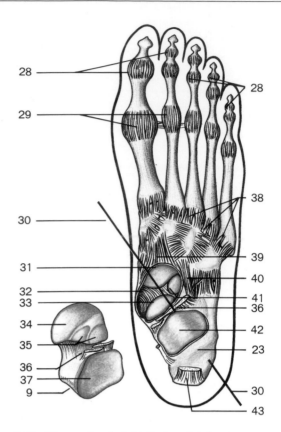

4.45 Bones of right foot (dorsal view).

4.46 Ligaments of right foot and talocalcaneonavicular joint (dorsal view). The talus has been rotated to show the corresponding articular surface of the three interconnected bones (schematic drawing).

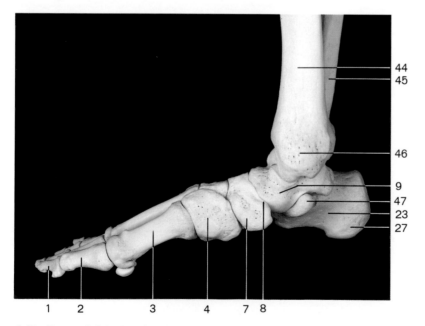

4.47 Bones of right foot (medial view).

1   Distal phalanx of great toe
2   Proximal phalanx of great toe
3   First metatarsal bone
   a   Head
   b   Base
4   Medial cuneiform bone
5   Intermediate cuneiform bone
6   **Cuneonavicular joint**
7   Navicular bone
8   **Talocalcaneonavicular joint**
9   Talus
10   Trochlea of talus
11   Posterior talar process
12   Distal phalanges
13   Middle phalanx
14   **Interphalangeal joints**
15   Proximal phalanges
16   **Metatarsophalangeal joints**
17   Metatarsal bones
18   **Tarsometatarsal joints**
19   Lateral cuneiform bone
20   Tuberosity of 5th metatarsal bone
21   Cuboid bone
22   **Calcaneocuboid joint**
23   Calcaneus
24   Tarsal sinus
25   Lateral malleolar surface of talus

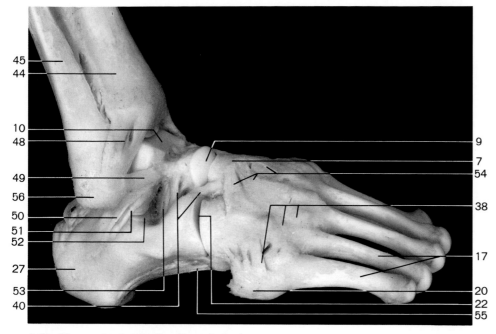

4.48   Ligaments of right foot (lateral view).

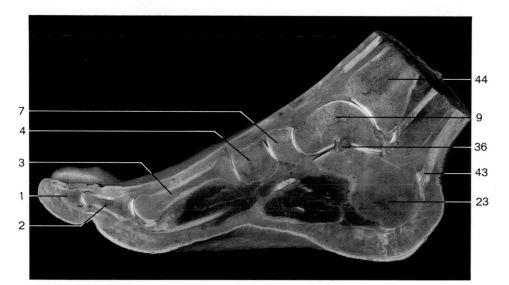

4.49   Longitudinal section of right foot through the great toe.

26	Peroneal trochlea of calcaneus	42	Posterior talar articular surface of calcaneus
27	Calcaneal tuberosity	43	Calcaneus tendon (Achilles tendon)
28	Articular capsules of interphalangeal joints	44	Tibia
29	Articular capsules of metatarsophalangeal joints	45	Fibula
30	Axis for inversion and eversion	46	Medial malleolus of tibia
31	Articular surface of navicular bone	47	Sustentaculum tali (of talus)
32	Plantar calcaneonavicular ligament	48	Anterior tibiofibular ligament (**talocrural joint**)
33	Middle talar articular surface of calcaneus	49	Anterior talofibular ligament
34	Navicular articular surface of talus	50	Calcaneofibular ligament
35	Anterior and middle calcaneal surfaces of talus	51	Lateral talocalcaneal ligament
36	Interosseous talocalcaneal ligament (cut)	52	**Subtalar joint**
37	Posterior calcaneal surface of talus	53	Interosseous talocalcaneal ligament
38	Dorsal tarsometatarsal ligaments	54	Dorsal cuneonavicular ligaments
39	Talonavicular ligament	55	Long plantar ligament
40	Bifurcated ligament	56	Lateral malleolus of fibula
41	Anterior talar articular surface of calcaneus		

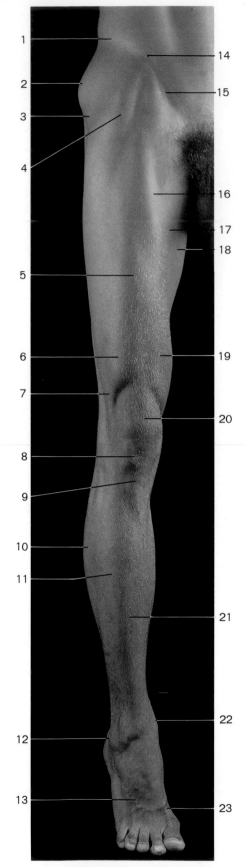

4.50   Surface anatomy of lower extremity (anterior view).

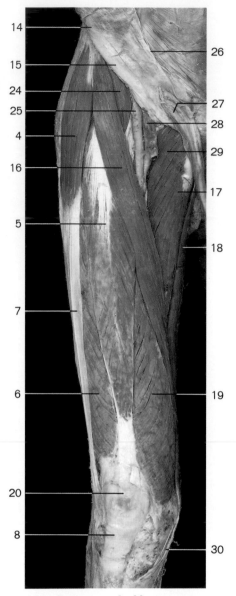

4.51   Extensor and adductor muscles of right thigh (anterior view).

 1   Iliac crest
 2   Gluteus medius
 3   Greater trochanter of femur
 4   Tensor muscle of fascia lata
 5   Rectus femoris
 6   Vastus lateralis
 7   Iliotibial tract
 8   Patellar ligament
 9   Tibial tuberosity
10   Gastrocnemius
11   Tibialis anterior
12   Lateral malleolus
13   Dorsal venous arch
14   Anterior superior iliac spine

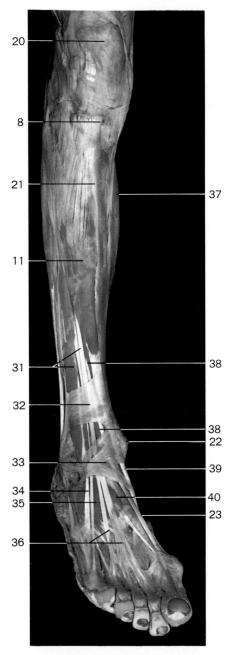

4.52 Extensor muscles of right leg and foot (oblique anterolateral view).

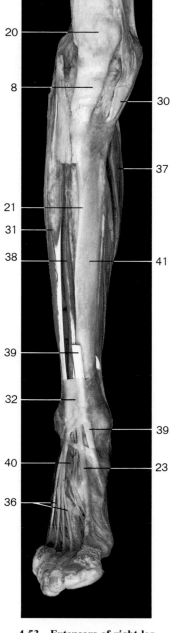

4.53 Extensors of right leg and foot (anterior view). Tibialis anterior has been cut.

4.54 Right leg and dorsum of foot, middle layer (oblique anterior view). The extensor digitorum longus has been cut and reflected laterally.

15 Inguinal ligament	28 Femoral vein	40 Extensor hallucis brevis
16 Sartorius	29 Pectineus	41 Tibia
17 Adductor longus	30 Common tendon of sartorius, gracilis and semitendinosus	42 Tendon of biceps femoris
18 Gracilis		43 Common peroneal nerve
19 Vastus medialis	31 Extensor digitorum longus	44 Muscular branches of deep peroneal nerve
20 Patella	32 Superior extensor retinaculum	
21 Anterior margin of tibia	33 Inferior extensor retinaculum	45 Superficial peroneal nerve
22 Medial malleolus	34 Tendon of peroneus tertius	46 Anterior tibial artery
23 Tendon of extensor hallucis longus	35 Extensor digitorum brevis	47 Deep peroneal nerve
24 Iliopsoas	36 Tendons of extensor digitorum longus	48 Dorsal artery of foot
25 Femoral artery	37 Soleus	49 Dorsal digital nerves (terminal branches of deep peroneal nerve)
26 Aponeurosis of external oblique muscle of abdomen	38 Extensor hallucis longus	
	39 Tendon of tibialis anterior	
27 Spermatic cord		

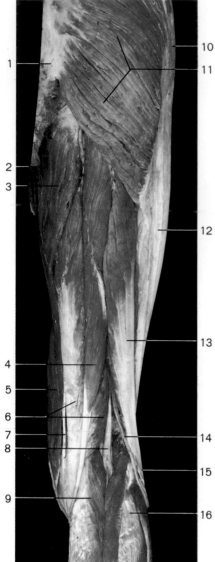

4.55 Dorsal muscles of right thigh, superficial layer (posterior view).

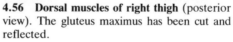

4.56 Dorsal muscles of right thigh (posterior view). The gluteus maximus has been cut and reflected.

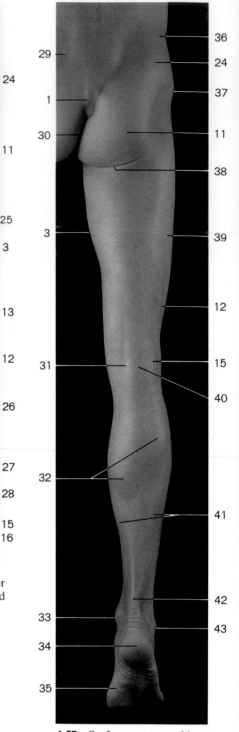

4.57 Surface anatomy of lower extremity (posterior view).

1	Coccyx	24	Gluteus medius
2	Anus	25	Adductor minimus
3	Adductor magnus	26	Short head of biceps femoris
4	Semitendinosus	27	Popliteal surface of femur
5	Sartorius	28	Plantaris
6	Semimembranosus	29	Posterior superior iliac spine
7	Tendon of gracilis	30	Natal cleft
8	Tibial nerve	31	Tendon of semitendinosus and semimembranosus
9	Medial head of gastrocnemius	32	Gastrocnemius
10	Tensor muscle of fascia lata	33	Medial malleolus
11	Gluteus maximus	34	Calcaneus
12	Iliotibial tract	35	Abductor hallucis
13	Long head of biceps femoris	36	Iliac crest
14	Common peroneal nerve	37	Greater trochanter of femur
15	Tendon of biceps femoris	38	Fold of buttock
16	Lateral head of gastrocnemius	39	Vastus lateralis
17	Piriformis	40	Popliteal fossa
18	Superior gemellus	41	Soleus
19	Obturator internus	42	Calcaneus tendon
20	Inferior gemellus	43	Lateral malleolus of fibula
21	Ischial tuberosity		
22	Quadratus femoris		
23	Semitendinosus muscle with intermediate tendon		

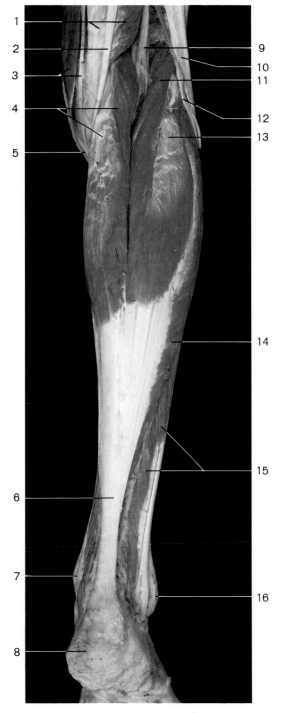

4.58 Flexor muscles of right leg (posterior view).

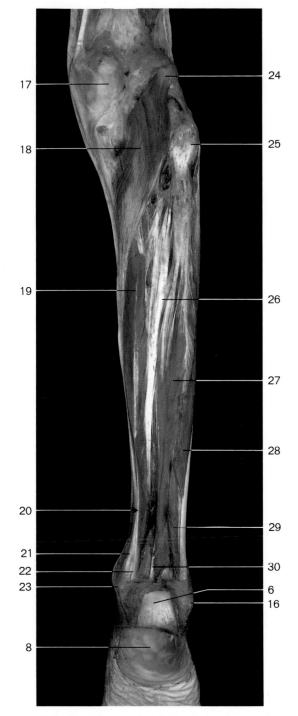

4.59 Deep flexors of right leg (posterior view).

1	Semimembranosus	11	Plantaris
2	Semitendinosus	12	Common peroneal nerve
3	Sartorius, tendon of gracilis	13	Lateral head of gastrocnemius
4	Medial head of gastrocnemius	14	Soleus
5	Common tendon of gracilis, sartorius and semitendinosus	15	Peroneus longus and brevis
6	Calcaneus tendon	16	Lateral malleolus
7	Medial malleolus	17	Medial condyle of femur
8	Calcaneal tuberosity	18	Popliteus
9	Tibial nerve	19	Flexor digitorum longus
10	Biceps femoris	20	Crossing of tendons in leg

21	Tendon of tibialis posterior
22	Tendon of flexor digitorum longus
23	Medial malleolus
24	Lateral condyle of femur
25	Head of fibula
26	Tibialis posterior
27	Flexor hallucis longus
28	Peroneus longus
29	Peroneus brevis
30	Tendon of flexor hallucis longus

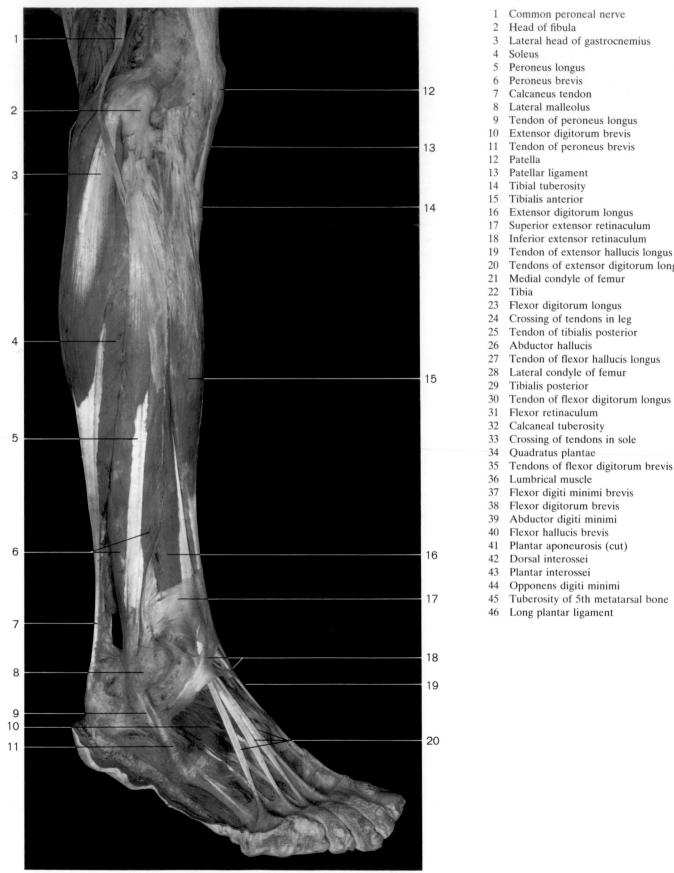

1   Common peroneal nerve
2   Head of fibula
3   Lateral head of gastrocnemius
4   Soleus
5   Peroneus longus
6   Peroneus brevis
7   Calcaneus tendon
8   Lateral malleolus
9   Tendon of peroneus longus
10   Extensor digitorum brevis
11   Tendon of peroneus brevis
12   Patella
13   Patellar ligament
14   Tibial tuberosity
15   Tibialis anterior
16   Extensor digitorum longus
17   Superior extensor retinaculum
18   Inferior extensor retinaculum
19   Tendon of extensor hallucis longus
20   Tendons of extensor digitorum longus
21   Medial condyle of femur
22   Tibia
23   Flexor digitorum longus
24   Crossing of tendons in leg
25   Tendon of tibialis posterior
26   Abductor hallucis
27   Tendon of flexor hallucis longus
28   Lateral condyle of femur
29   Tibialis posterior
30   Tendon of flexor digitorum longus
31   Flexor retinaculum
32   Calcaneal tuberosity
33   Crossing of tendons in sole
34   Quadratus plantae
35   Tendons of flexor digitorum brevis
36   Lumbrical muscle
37   Flexor digiti minimi brevis
38   Flexor digitorum brevis
39   Abductor digiti minimi
40   Flexor hallucis brevis
41   Plantar aponeurosis (cut)
42   Dorsal interossei
43   Plantar interossei
44   Opponens digiti minimi
45   Tuberosity of 5th metatarsal bone
46   Long plantar ligament

4.60   Muscles of leg and foot (right leg, lateral view).

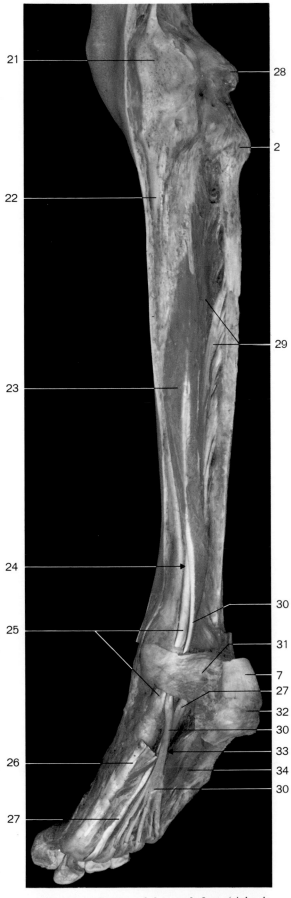

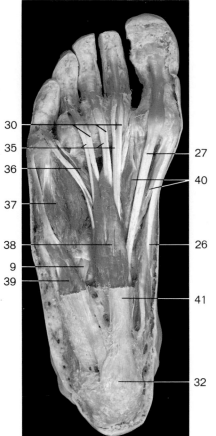

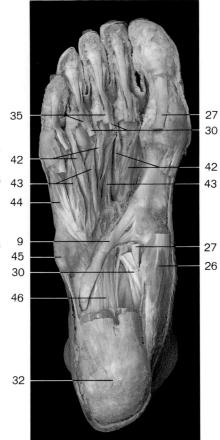

4.62 Muscles of sole of foot, superficial layer (viewed from below). The plantar aponeurosis and the fasciae of the superficial muscles have been removed.

4.63 Muscles of sole of foot, deepest layer (viewed from below). The interosseous muscles and the canal for the tendon of peroneus longus are displayed.

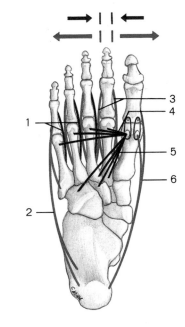

1   Plantar interossei (black)
2   Abductor digiti minimi (red)
3   Dorsal interossei (red)
4   Transverse head of adductor hallucis (black)
5   Oblique head of adductor hallucis (black)
6   Abductor hallucis (red)

4.64 Course of abductor and adductor muscles of foot (schematic drawing).
Red arrows = abduction;
black arrows = adduction.

4.61 Deep flexors of leg and foot (right leg, oblique posteromedial view). Flexor digitorum brevis and flexor hallucis longus have been removed.

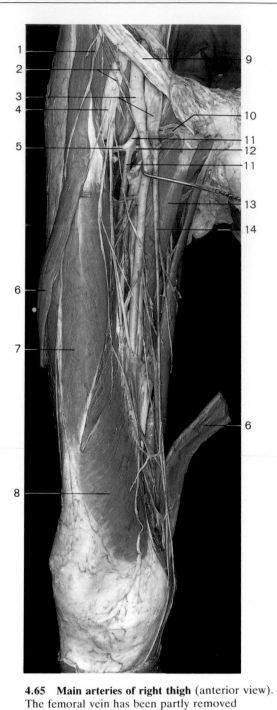

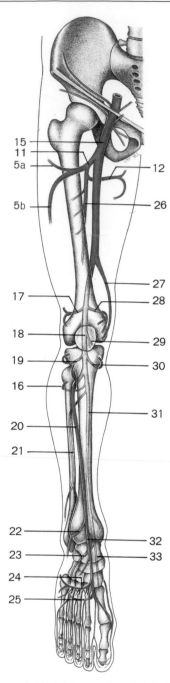

4.65   Main arteries of right thigh (anterior view).
The femoral vein has been partly removed
to show the deep femoral artery.

4.66   Main arteries of right lower extremity,
anterior view (schematic drawing).

1 Lateral cutaneous femoral nerve	11 Deep femoral artery	24 Arcuate artery with dorsal metatarsal
2 Superficial and deep iliac circumflex	12 Medial circumflex femoral artery	arteries
arteries	13 Adductor longus	25 Plantar arch with plantar metatarsal
3 Femoral artery and vein	14 Great saphenous vein	arteries
4 Femoral nerve	15 Femoral artery	26 Deep femoral artery with perfo-
5 Lateral circumflex femoral artery	16 Posterior tibial recurrent artery	rating arteries
a Ascending branch	17 Lateral superior genicular artery	27 Descending genicular artery
b Descending branch	18 Popliteal artery	28 Medial superior genicular artery
6 Sartorius muscle (cut)	19 Lateral inferior genicular artery	29 Middle genicular artery
7 Rectus femoris	20 Anterior tibial artery	30 Medial inferior genicular artery
8 Vastus medialis	21 Peroneal artery	31 Posterior tibial artery
9 Inguinal ligament	22 Lateral anterior malleolar artery	32 Dorsalis pedis artery
10 External pudendal artery and vein	23 Lateral plantar artery	33 Medial plantar artery

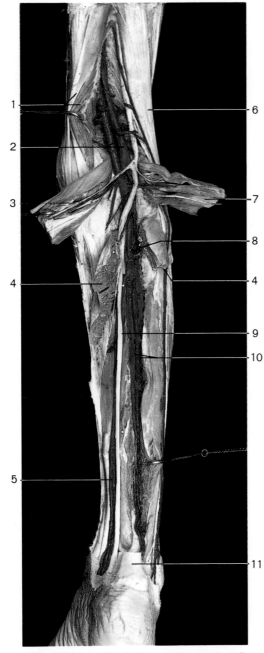

4.67 Right popliteal fossa and arteries of the leg (posterior view).

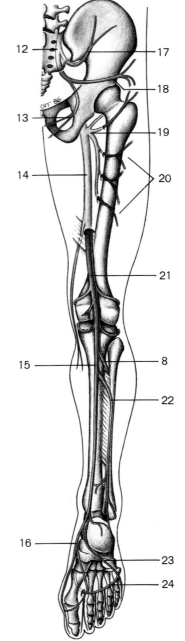

4.68 Arteries of right lower extremity, posterior view (schematic drawing).

1 Semitendinosus, semimembranosus	9 Tibial nerve	17 Superior gluteal artery
2 Popliteal artery and vein	10 Peroneal artery and vein	18 Inferior gluteal artery
3 Medial head of gastrocnemius (cut)	11 Calcaneus tendon (cut)	19 Deep femoral artery
4 Soleus	12 Internal iliac artery	20 Perforating arteries
5 Posterior tibial artery and vein	13 Obturator artery	21 Popliteal artery
6 Biceps femoris	14 Femoral artery	22 Peroneal artery
7 Lateral head of gastrocnemius (cut)	15 Posterior tibial artery	23 Lateral plantar artery
8 Anterior tibial artery	16 Medial plantar artery	24 Plantar arch

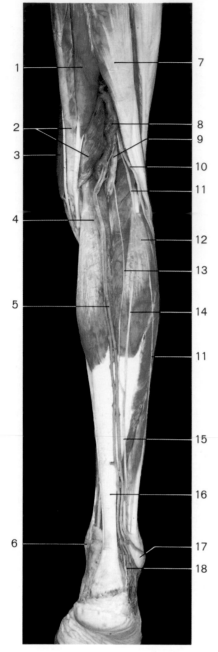

4.69 Cutaneous nerves and veins of right leg and foot (posterior view). The crural fascia and fasciae of the muscles have been removed.

4.70 Superficial veins of right lower extremity in medial anterior view (A) and posterior view (B). The veins have been injected with blue solution.

1	Semitendinosus	9	Tibial nerve	16	Calcaneus tendon
2	Semimembranosus	10	Common peroneal nerve	17	Lateral malleolus
3	Sartorius	11	Lateral cutaneous sural nerve	18	Lateral calcaneal branches of sural
4	Medial head of gastrocnemius	12	Lateral head of gastrocnemius		nerve
5	Small saphenous vein	13	Medial cutaneous sural nerve	19	Saphenous opening with femoral
6	Medial malleolus	14	Communicating branch of peroneal		artery and vein
7	Biceps femoris		nerve	20	Dorsal venous arch of foot
8	Varicosity of vein (pathological)	15	Sural nerve	21	Great saphenous vein

Saphenous veins are cutaneous veins, without accompanying arteries, that drain into the femoral vein. Congestion may occur in the lower extremity, due to low venous pressure and the force of gravity, leading to varicose veins.

The sciatic nerve, the largest and longest nerve in the body, extends more than 1 m to supply the external coxal muscle, peroneal muscle and muscles and skin of the back of the lower extremity.

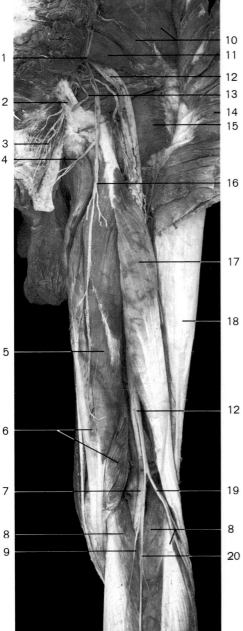

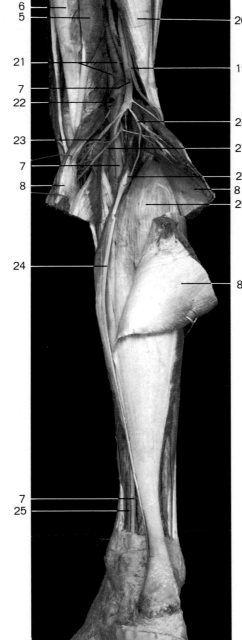

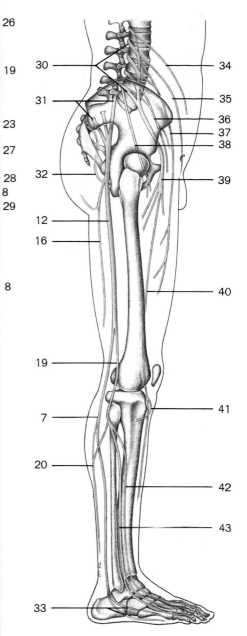

4.71 Posterior femoral region and gluteal region (right side, posterior view). The gluteus maximus has been cut and reflected.

4.72 Right popliteal fossa (middle layer) and leg (posterior view). The cutaneous veins and nerves have been removed and the gastrocnemius cut.

4.73 Nerves of right lower limb, lateral view (schematic drawing).

1	Inferior gluteal nerve	15	Quadratus femoris
2	Sacrotuberal ligament	16	Posterior femoral cutaneous nerve
3	Inferior rectal branches of pudendal nerve	17	Long head of biceps femoris
4	Perineal branch of posterior femoral cutaneous nerve	18	Iliotibial tract
5	Semitendinosus	19	**Common peroneal nerve**
6	Semimembranosus	20	Lateral cutaneous sural nerve
7	**Tibial nerve**	21	Popliteal artery and vein
8	Gastrocnemius	22	Small saphenous vein (cut)
9	Medial sural cutaneous nerve	23	Muscular branch of tibial nerve
10	Gluteus medius	24	Tendon of plantaris
11	Piriformis	25	Posterior tibial artery
12	**Sciatic nerve**	26	Biceps femoris
13	Inferior gluteal artery	27	Sural arteries
14	Gluteus maximus	28	Plantaris
		29	Soleus

30	Lumbar plexus (T12 ; L1–L4)
31	Sacral plexus (L4–L5; S1–S4)
32	**Pudendal nerve**
33	Medial and lateral plantar nerve
34	Iliohypogastric nerve
35	Ilioinguinal nerve
36	Genitofemoral nerve
37	Lateral cutaneous femoral nerve
38	Obturator nerve
39	**Femoral nerve**
40	Saphenous nerve
41	Infrapatellar branch of saphenous nerve
42	**Deep peroneal nerve**
43	**Superficial peroneal nerve**

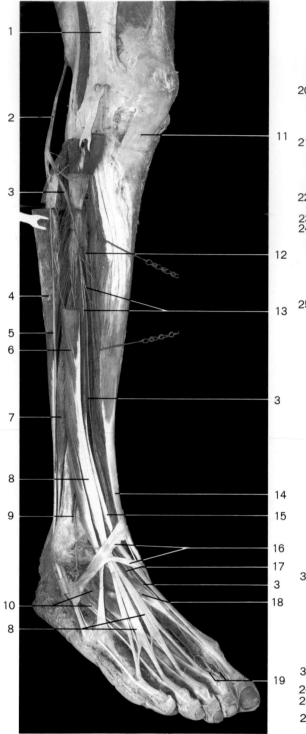

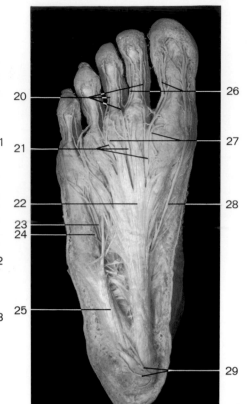

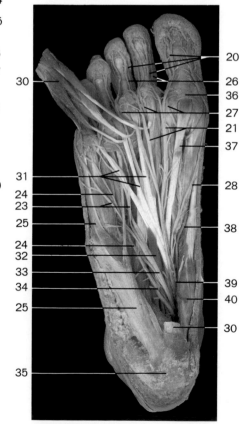

1   Tendon of biceps femoris
2   **Common peroneal nerve**
3   **Deep peroneal nerve**
4   Peroneus longus (cut)
5   Superficial peroneal nerve
6   Extensor digitorum longus (cut)
7   Peroneus brevis
8   Tendon of extensor digitorum longus
9   Lateral anterior malleolar artery
10 Extensor digitorum brevis
11 Patellar ligament
12 **Anterior tibial artery**
13 Muscular branches of deep peroneal nerve
14 Tendon of tibialis anterior
15 Extensor hallucis longus
16 Extensor retinaculum
17 **Dorsal artery of foot**
18 Extensor hallucis brevis
19 Dorsal digital nerves (terminal branches of deep peroneal nerves)
20 Proper plantar digital nerves
21 Common plantar digital nerves
22 Plantar aponeurosis
23 Superficial branch of lateral plantar nerve
24 Superficial branch of lateral plantar artery
25 Abductor digiti minimi
26 Proper plantar digital arteries
27 Common plantar digital arteries
28 Digital branch of medial plantar nerve to great toe
29 Medial calcaneal branches
30 Flexor digitorum brevis (cut)
31 Tendons of flexor digitorum longus
32 Quadratus plantae
33 **Lateral plantar nerve**
34 **Lateral plantar artery**
35 Calcaneal tuberosity
36 Digital synovial sheath
37 Tendon of flexor hallucis longus
38 **Medial plantar artery**
39 **Medial plantar nerve**
40 Abductor hallucis

4.75 Sole of right foot, superficial layer (inferior view); dissection of cutaneous nerves and vessels.

4.74 Right leg and dorsum of foot, deep layer (oblique anterior view). The extensor digitorum longus and peroneus longus have been cut or removed. The common peroneal nerve has been elevated to show its course around the head of the fibula.

4.76 Sole of right foot, middle layer (inferior view); dissection of vessels and nerves. The flexor digitorum brevis has been cut and reflected anteriorly.

5. Thoracic Organs

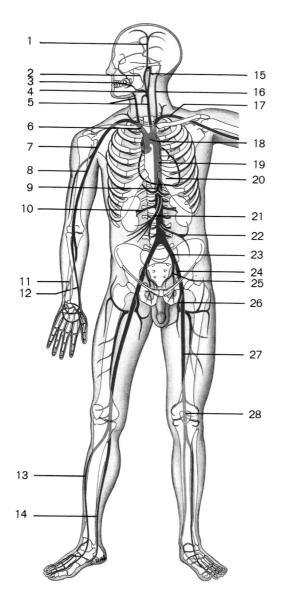

5.1 Main arteries of the human body (schematic drawing).

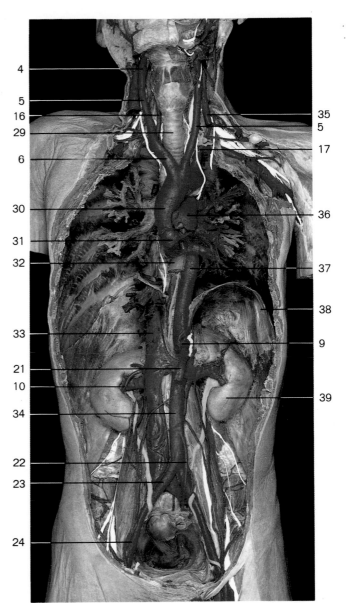

5.2 Major vessels of the trunk. The digestive organs and heart have been removed.
Red = arteries, blue = veins, green = lymphatic vessels, white = nerves, yellow = sympathetic nerve, orange = ureter

1   Superficial temporal artery
2   Maxillary artery
3   Facial artery
4   External carotid artery
5   Right and left common carotid arteries
6   Brachiocephalic trunk
7   Axillary artery
8   Brachial artery
9   Celiac trunk
10 Renal artery
11 Ulnar artery
12 Radial artery
13 Anterior tibial artery

14 Posterior tibial artery
15 Internal carotid artery
16 Vertebral artery
17 Left subclavian artery
18 Aortic arch
19 Descending aorta
20 Position of heart
21 Superior mesenteric artery
22 Inferior mesenteric artery
23 Common iliac artery
24 External iliac artery
25 Internal iliac artery
26 Deep femoral artery
27 Femoral artery

28 Popliteal artery
29 Trachea
30 Ascending aorta
31 Aortic semilunar valve
32 Esophagus (cut)
33 Inferior vena cava
34 Abdominal aorta
35 Internal jugular vein
36 Pulmonary trunk and valve
37 Thoracic aorta
38 Diaphragm
39 Kidney

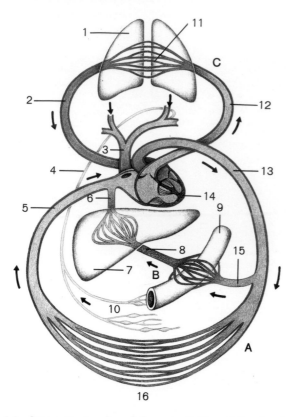

5.3   Schematic drawing of the organization of the circulatory system showing the systemic and pulmonary circulations and lymphatic vessels. Systemic arteries and pulmonary veins carry oxygenated blood (red), systemic veins and pulmonary arteries carry deoxygenated blood (blue), and lymphatic vessels carry lymph (yellow).
Arrows show direction of flow.
A = Systemic circulation
B = Hepatic portal circulation
C = Pulmonary circulation

1   Lung
2   Pulmonary vein
3   Superior vena cava
4   Thoracic duct
5   Inferior vena cava
6   Hepatic vein
7   Liver
8   Hepatic portal vein
9   Small intestine with capillary network
10 Lymphatic vessels and lymph nodes
11 Pulmonary capillary network
12 Pulmonary artery
13 Aorta
14 Heart
15 Mesenteric artery
16 Systemic capillary network
17 Internal jugular vein, right common carotid artery
18 Brachiocephalic veins
19 Foramen ovale
20 Ductus venosus
21 Umbilical vein
22 Colon
23 Umbilical arteries
24 Trachea
25 Aortic arch
26 Left pulmonary artery
27 Ductus arteriosus (Bottalo's duct)
28 Small intestine
29 Umbilicus
30 Urachus and urinary bladder
31 Placenta

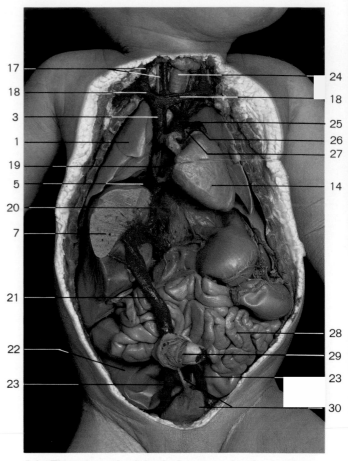

5.4   Thoracic and abdominal organs in the newborn (anterior view). The main vessels are colored to show the fetal circulatory system. The right atrium has been opened to show the foramen ovale. The left lobe of the liver has been removed.

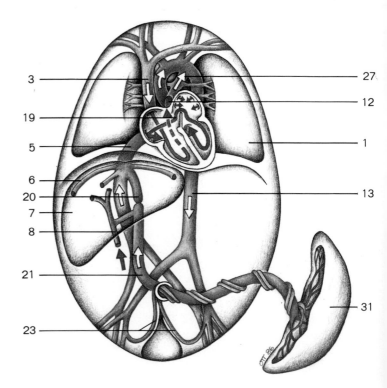

5.5   Schematic drawing of fetal circulation. Arrows indicate direction of blood flow. Blood vessels are colored to show the degree of oxygenation of the blood.
Red = oxygenated,   blue = deoxygenated,   purple = mixed.

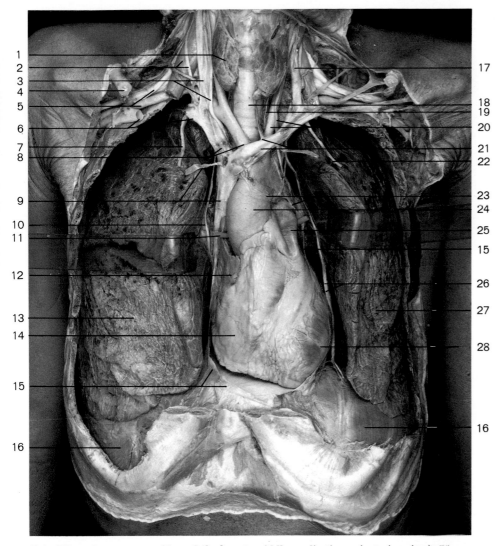

1	Thyroid gland
2	Phrenic nerve and scalenus anterior
3	Vagus nerve, internal jugular vein
4	Clavicle (cut)
5	Brachial plexus and subclavian artery
6	Right subclavian vein (cut)
7	Right inrernal thoracic artery
8	Brachiocephalic trunk and right brachiocephalic vein
9	Superior vena cava and thymic vein
10	Right phrenic nerve
11	Transverse pericardial sinus (probe)
12	Right auricle
13	Middle lobe of right lung
14	Right ventricle
15	Cut edge of pericardium
16	Diaphragm
17	Internal jugular vein
18	Trachea
19	Recurrent laryngeal nerve
20	Left common carotid artery and vagus nerve
21	Left brachiocephalic vein and inferior thyroid vein
22	Left internal thoracic artery (cut)
23	Upper margin of pericardial sac
24	Ascending aorta
25	Pulmonary trunk
26	Left phrenic nerve and left pericardiacophrenic artery and vein
27	Upper lobe of left lung
28	Left ventricle

5.6 Thoracic organs, position of the heart, middle mediastinum (anterior view). The anterior wall of the thorax, the costal pleura and the pericardium have been removed and the lungs slightly reflected.

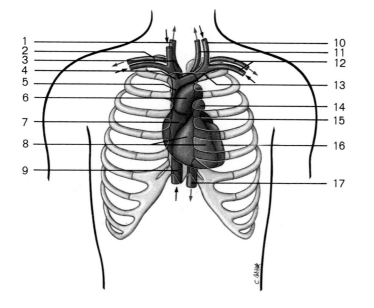

1	Right common carotid artery
2	Right subclavian artery
3	Brachiocephalic trunk
4	Right brachiocephalic vein
5	Superior vena cava
6	Ascending aorta
7	Right atrium
8	Right ventricle
9	Inferior vena cava
10	Left internal jugular vein
11	Left common carotid artery
12	Left axillary artery and vein
13	Left brachiocephalic vein
14	Pulmonary trunk
15	Left atrium
16	Left ventricle
17	Descending aorta

5.7 Location of the heart and proximal vessels within the thorax (schematic drawing).

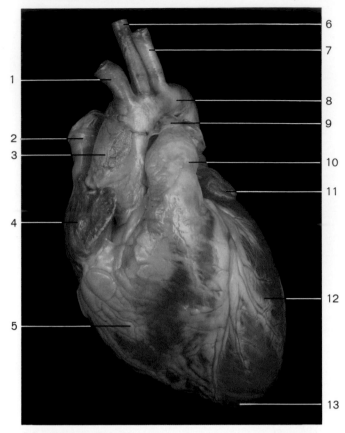

5.8 Heart in diastole, anterior view. The ventricles are relaxed, atria are contracted.

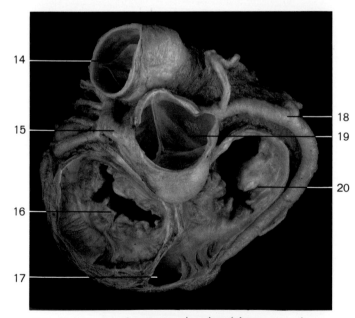

5.9 Valves of the heart, superior view (above : anterior wall of the heart).

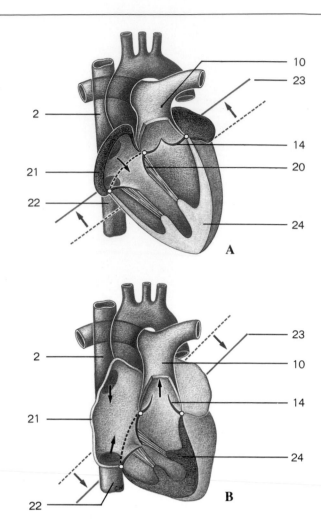

5.10 Morphological changes during heart movements (schematic drawing after Gauer). Note the changes in position of the valves. Contracted portions of the heart are indicated in dark gray.

A. **Diastole**, muscles of ventricles relaxed; atrioventricular valves open, semilunar valves closed.

B. **Systole**, muscles of ventricles contracted; atrioventricular valves closed, semilunar valves open.

1 Brachiocephalic trunk	10 Pulmonary trunk	18 Right coronary artery
2 Superior vena cava	11 Left auricle	19 Aortic semilunar valve
3 Ascending aorta	12 Left ventricle	20 Tricuspid valve (right
4 Right auricle	13 Apex of heart	atrioventricular valve)
5 Right ventricle	14 Pulmonary semilunar valve	21 Myocardium of right atrium
6 Left common carotid artery	15 Left coronary artery	22 Inferior vena cava
7 Left subclavian artery	16 Bicuspid (mitral) valve (left	23 Position of valves
8 Aortic arch	atrioventricular valve)	24 Myocardium of right ventricle
9 Ligamentum arteriosum	17 Coronary sinus	

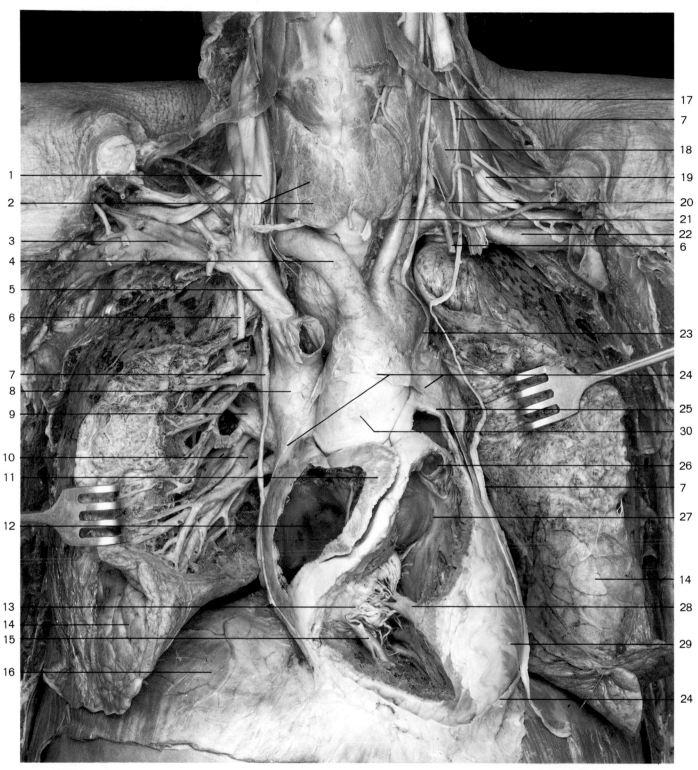

5.11 Heart with valves *in situ* (anterior view). Anterior wall of thorax, pleura and anterior portion of pericardium have been removed. The right atrium and ventricle have been opened to show the atrioventricular and pulmonary valves.

1	Internal jugular vein	11	Right auricle	20	Thyrocervical trunk
2	Thyroid gland	12	Right atrium	21	Left common carotid artery
3	Right subclavian vein	13	Tricuspid valve (anterior cusp)	22	Left subclavian artery
4	Brachiocephalic trunk		with chordae tendineae	23	Left recurrent laryngeal nerve
5	Right brachiocephalic vein	14	Lung	24	Cut edge of pericardium
6	Internal thoracic arteries	15	Posterior papillary muscle	25	Pulmonary trunk (opened)
7	Phrenic nerves	16	Diaphragm	26	Pulmonary semilunar valve
8	Superior vena cava	17	Left vagus nerve	27	Arterial cone of right ventricle
9	Right pulmonary vein	18	Anterior scalenus muscle	28	Anterior papillary muscle
10	Branch of pulmonary artery	19	Brachial plexus	29	Left ventricle
				30	Ascending aorta

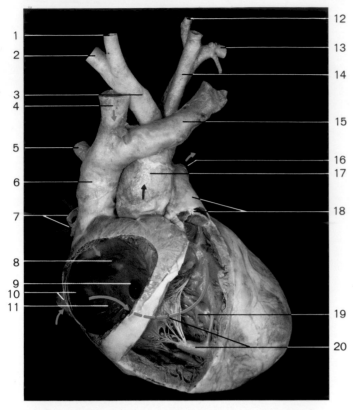

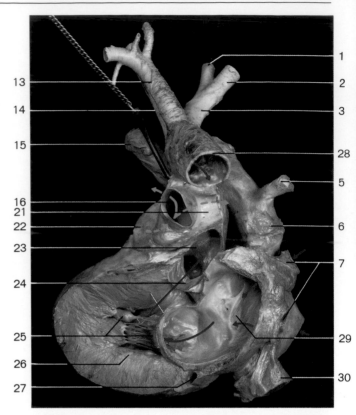

5.12A **Blood circulation within the heart** (anterior view). The anterior wall of the right atrium and ventricle has been partially removed. The pulmonary trunk has been opened to show the pulmonary semilunar valves. Arrows indicate direction of flow.

5.12B **Blood circulation within the heart** (posterior view). The posterior wall of the left atrium and ventricle has been removed. The aortic bulb has been opened. Arrows indicate direction of blood flow.

1   Right common carotid artery
2   Right subclavian artery
3   Brachiocephalic trunk
4   Right brachiocephalic vein
5   Azygos vein
6   Superior vena cava
7   Right pulmonary veins
8   Fossa ovalis
9   Coronary sinus valve
10 Opening of inferior vena cava
11 **Right atrium**
12 Left vertebral artery
13 Left subclavian artery
14 Left common carotid artery
15 Left brachiocephalic vein
16 Left pulmonary artery
17 Ascending aorta
18 Pulmonary trunk and valve
19 **Right ventricle**
20 Tricuspid valve and anterior papillary muscle
21 Right pulmonary artery
22 Pulmonary trunk
23 Aortic bulb
24 Aortic semilunar valve
25 Bicuspid (mitral) valve and posterior papillary muscle
26 **Left ventricle**
27 Coronary sinus
28 Aortic arch
29 **Left atrium**
30 Inferior vena cava

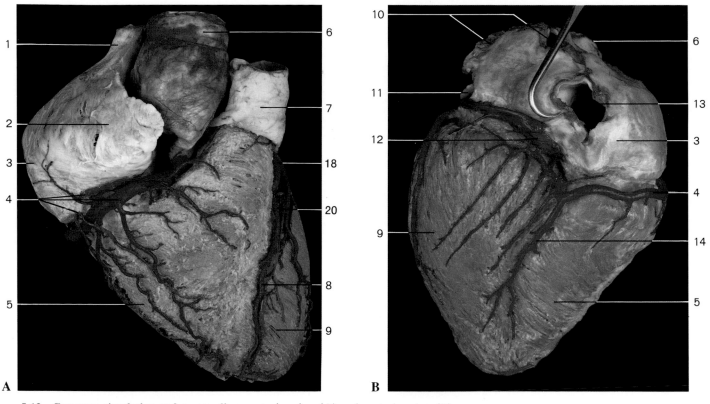

5.13 Coronary circulation and myocardium, anterior view (A) and posterior view (B).

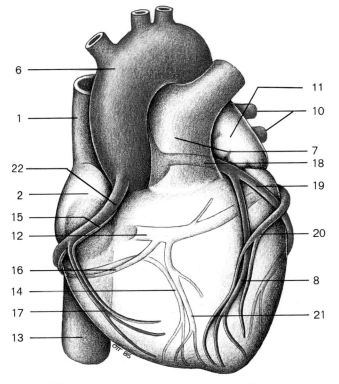

5.14 Schematic drawing of major cardiac vessels (anterior view). Vessels on posterior surface appear as lighter colors.

1   Superior vena cava
2   Right auricle
3   Right atrium
4   Right coronary artery and anterior cardiac veins
5   Right ventricle
6   Aorta
7   Pulmonary trunk
8   Anterior interventricular branch of left coronary artery
9   Left ventricle
10 Pulmonary veins
11 Left auricle
12 Coronary sinus
13 Inferior vena cava
14 Posterior interventricular branch of right coronary artery
15 Anterior cardiac vein
16 Small cardiac vein
17 Right marginal branch of right coronary artery
18 Left coronary artery
19 Circumflex artery
20 Great cardiac vein
21 Middle cardiac vein
22 Right coronary artery

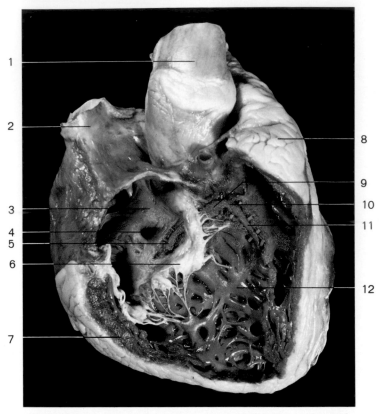

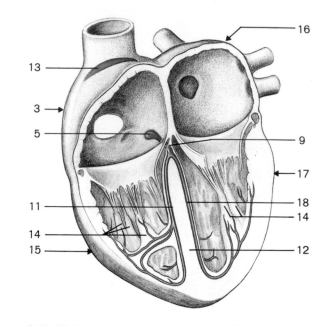

5.15 Right ventricle, dissection of **atrioventricular node, atrioventricular bundle (bundle of His)** and right limb of bundle branch (probes).

5.16 Excitatory and conducting system of the heart (schematic drawing).

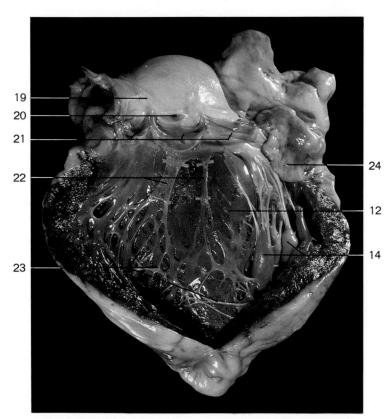

5.17 Left ventricle, dissection of the left limb of bundle branch of conducting system (probes).

1   Ascending aorta
2   Superior vena cava
3   Right atrium
4   Opening of coronary sinus
5   Atrioventricular node
6   Septal cusp of tricuspid valve
7   Wall of right ventricle
8   Pulmonary trunk
9   Atrioventricular bundle (bundle of His)
10   Bifurcation of atrioventricular bundle
11   Right bundle branch
12   Interventricular septum
13   Sinoatrial node
14   Papillary muscles
15   Right ventricle
16   Left atrium
17   Left ventricle
18   Left bundle branch
19   Aortic sinus
20   Opening to left coronary artery
21   Aortic semilunar valve
22   Branches of left bundle branch
23   Purkinje fibers
24   Left auricle

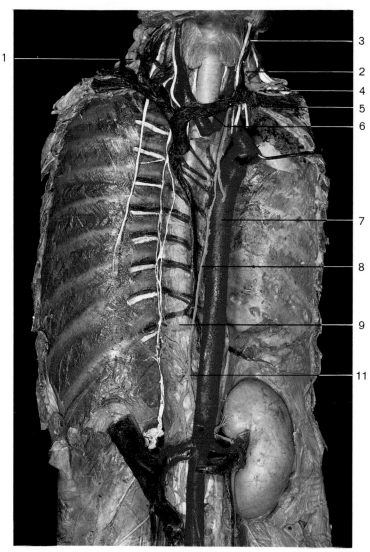

5.18 Major lymphatic vessels of the trunk (anterior view).
Green = lymphatic vessels and lymph nodes;
white = nerves.

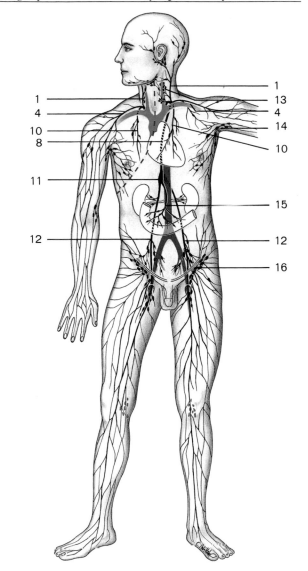

5.19 Lymphatic system (schematic drawing). Course of the main lymphatic vessels and lymph nodes in the body.

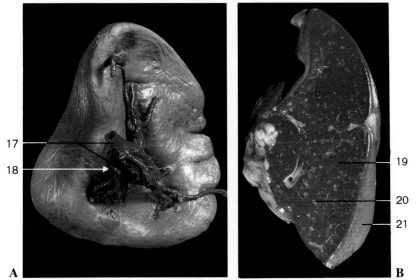

5.20 Spleen. Visceral surface (A) shows splenic vessels; cut surface (B) shows white and red pulp.

1	Jugular trunk	11 Cisterna chyli
2	Deep cervical lymph nodes	12 Lumbar trunk
3	Internal jugular vein	13 **Cervical lymph nodes**
4	Subclavian trunk	14 **Axillary lymph nodes**
5	Left subclavian vein	15 Intestinal trunk
6	Left brachiocephalic vein	16 **Inguinal lymph nodes**
7	Thoracic aorta	17 Splenic artery and vein
8	**Thoracic duct**	18 Hilum of spleen
9	Trunk of mediastinal lymph nodes	19 Red pulp
10	Bronchiomediastinal trunk	20 White pulp
		21 Fibrous capsule

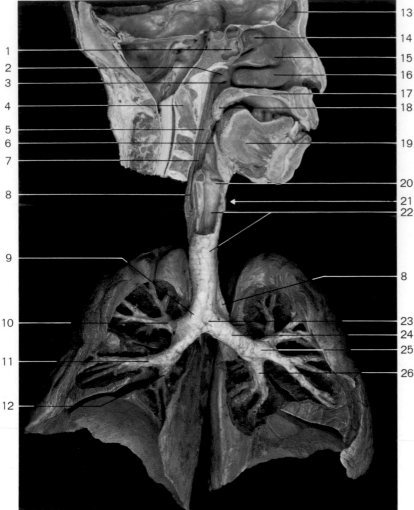

1	Sphenoidal sinus
2	Nasopharynx
3	Pharyngeal opening of auditory tube
4	Dens of axis
5	Oropharynx (isthmus of fauces)
6	Epiglottis
7	Laryngopharynx
8	Esophagus
9	Right primary (main) bronchus
10	Right superior lobar bronchus ⎫ Secondary
11	Right middle lobar bronchus ⎬ bronchi
12	Right inferior lobar bronchus ⎭
13	Frontal sinus
14	Superior nasal concha
15	Middle nasal concha
16	Inferior nasal concha
17	Hard palate
18	Soft palate
19	Tongue
20	Vocal fold
21	Larynx
22	Trachea
23	Bifurcation of trachea
24	Left primary (main) bronchus
25	Left superior lobar bronchus ⎫ Secondary
26	Left inferior lobar bronchus ⎬ bronchi
27	Nasal cavity
28	Pharynx
29	Superior lobe of right lung
30	Middle lobe of right lung
31	Inferior lobe of right lung
32	Costal arch
33	Superior lobe of left lung
34	Secondary bronchi
35	Tertiary bronchi of bronchopulmonary segments
36	Inferior lobe of left lung

5.21   Respiratory system. The lungs have been fixed in expiration.

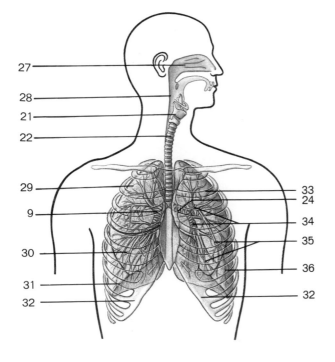

5.22   Organization and location of respiratory organs (schematic drawing).

1 Apex of lung
2 Superior lobe
3 Horizontal fissure of right lung
4 Oblique fissure
5 Middle lobe of right lung
6 Inferior lobe
7 Impression of rib
8 Groove for subclavian artery
9 Groove for azygos vein
10 Branches of right pulmonary artery
11 Bronchi
12 Right pulmonary veins
13 Pulmonary ligament
14 Diaphragmatic surface
15 Groove for aortic arch
16 Left pulmonary artery
17 Branches of left pulmonary veins
18 Left primary bronchus
19 Groove for thoracic aorta
20 Groove for esophagus
21 Cardiac impression
22 Bronchiole
23 Bronchial artery
24 Cartilage and smooth muscle
25 Terminal bronchioles
26 Respiratory bronchiole
27 Tributary of pulmonary vein
28 Interlobular septum
29 Pulmonary pleura
30 Branch of pulmonary artery
31 Alveolar duct
32 Alveolar sac
33 Pulmonary capillaries
34 Alveoli of lung
35 Lingula of lung

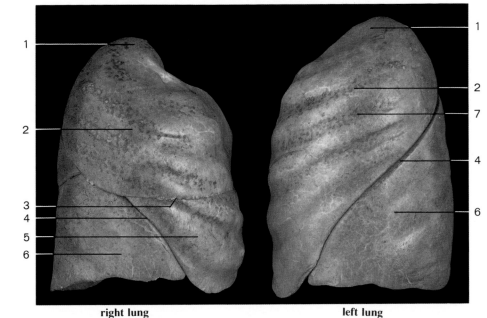

right lung   left lung

5.23 Lateral view of the lungs.

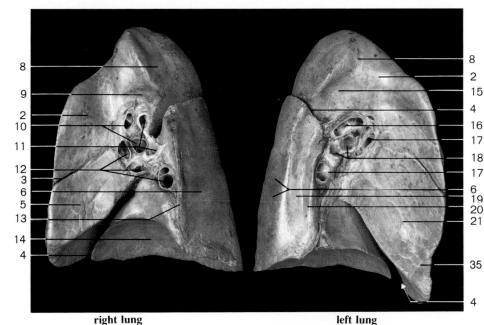

right lung   left lung

5.24 Medial view of the lungs.

5.25 Terminal portion of respiratory tract (schematic drawing). Arrows = directions for air and blood flow.

5.26 Cast of pulmonary alveoli.

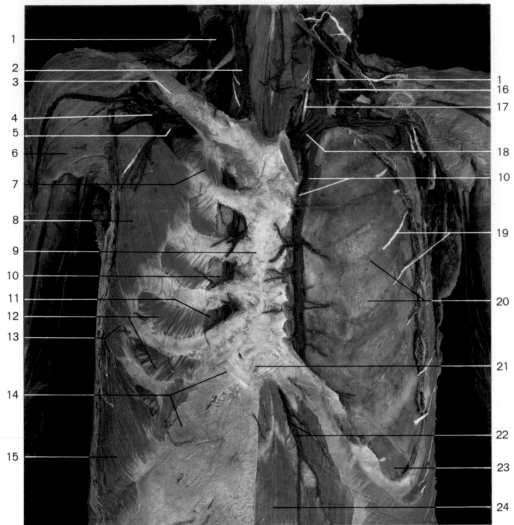

1	Internal jugular vein
2	Anterior jugular vein
3	Clavicle
4	Thoracoacromial artery
5	Right axillary vein
6	Pectoralis major
7	External intercostal muscle
8	Pectoralis minor
9	Body of sternum
10	Internal thoracic artery and vein
11	Fascicles of tranversus thoracis
12	Internal intercostal muscles
13	Serratus anterior
14	Costal arch
15	External oblique muscle
16	Transverse cervical artery
17	Vagus nerve
18	Left brachiocephalic vein
19	Intercostal nerves, arteries and veins
20	Costal pleura (parietal pleura)
21	Xiphoid process
22	Superior epigastric artery and vein
23	Diaphragm
24	Rectus abdominis
25	Larynx
26	Trachea
27	Lung covered with visceral pleura
28	Diaphragmatic pleura (parietal pleura)
29	Cupula of pleura
30	Costodiaphragmatic recess
31	Liver
32	Thyroid gland
33	Thymus gland
34	Heart and pericardium
35	Stomach
36	Mediastinal pleura (parietal pleura)

5.27   Thoracic wall (anterior view). The left clavicle and ribs have been partially removed, and the right intercostal spaces opened to show the internal thoracic artery and vein. Green = lymph nodes (at right intercostal space) and thoracic duct (at left venous angle)

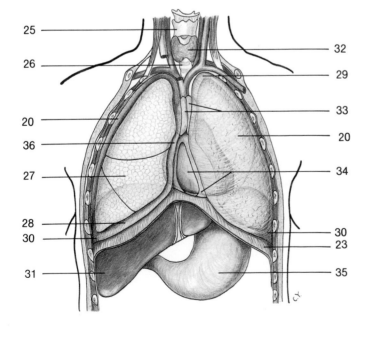

5.28   Location of thoracic and upper abdominal organs and pleurae (schematic drawing). Anterior wall of the pericardium has been opened to show the heart. On the right side the anterior part of the costal pleura has been removed to show the lung.

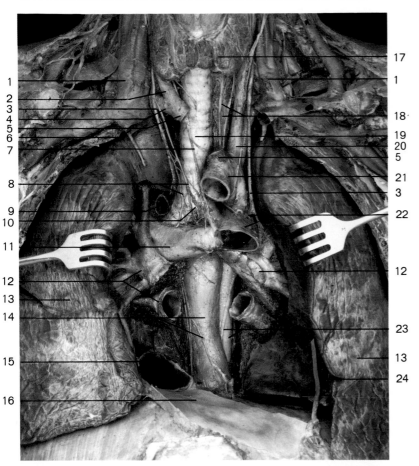

5.29 Mediastinal organs exposed after removal of heart and pericardium (anterior view). Both lungs have been slightly reflected.

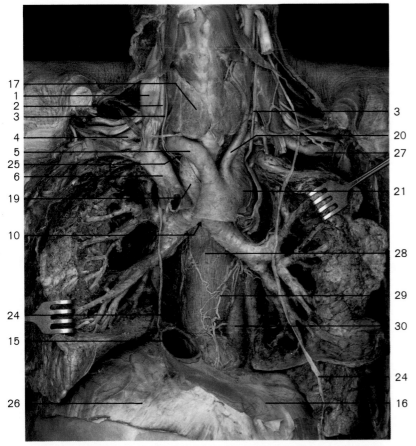

5.30 Bronchial tree *in situ* (anterior view). Heart and pericardium have been removed. Tertiary bronchi of bronchopulmonary segments are dissected.

1   Internal jugular vein
2   Right common carotid artery
3   Vagus nerve
4   Right subclavian artery
5   Brachiocephalic trunk
6   Right brachiocephalic vein
7   Superior cervical cardiac branch of vagus nerve
8   Inferior cervical cardiac branches of vagus nerve
9   Azygos vein (cut)
10 Bifurcation of trachea
11 Right pulmonary artery
12 Pulmonary veins (cut)
13 Lung
14 Esophagus and branches of vagus nerve
15 Inferior vena cava (cut)
16 Pericardium
17 Thyroid gland
18 Esophagus and left recurrent laryngeal nerve
19 Trachea
20 Left common carotid artery
21 Aortic arch (cut)
22 Pulmonary trunk and left recurrent laryngeal nerve branching off vagus nerve
23 Thoracic aorta and vagus nerve
24 Phrenic nerve
25 Right recurrent laryngeal nerve
26 Diaphragm
27 Left subclavian artery
28 Esophagus
29 Esophageal plexus
30 Thoracic aorta
31 Tracheal lymph nodes
32 Superior tracheobronchial lymph nodes
33 Left pulmonary artery
34 Bronchopulmonary lymph nodes

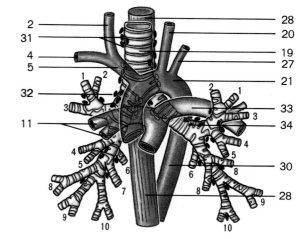

5.31 Anatomical relation of aorta, pulmonary trunk and esophagus to trachea and bronchial tree (schematic drawing). Bronchi of bronchopulmonary segments are numbered 1-10.

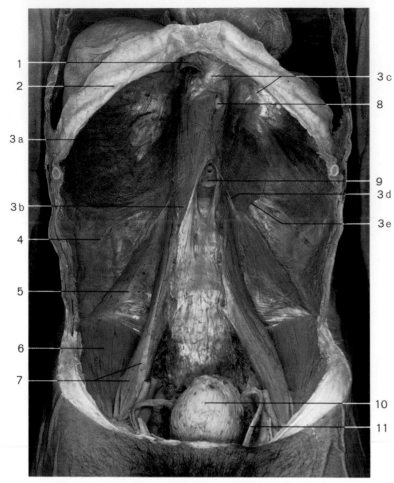

1 Inferior vena cava (cut)
2 Costal arch
3 Diaphragm
   a Costal part
   b Right crus of lumbar part
   c Central tendon
   d Medial lumbocostal arch
   e Lateral lumbocostal arch
4 Transversalis fascia
5 Quadratus lumborum
6 Iliacus
7 Psoas major and minor
8 Esophagus
9 Abdominal aorta (cut)
10 Urinary bladder
11 Ductus deferens

5.32 Diaphragm and posterior abdominal wall (anterior view). Abdominal and pelvic organs, except the urinary bladder, have been removed. The costal arch remains in place.

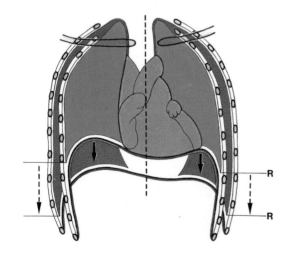

5.33 Schematic drawing of thoracic volumes (anterior view) showing changes in the position of the diaphragm and thoracic cage. During inspiration the diaphragm moves downwards and the lower part of the thoracic cage expands forwards and laterally (arrows), causing the costodiaphragmatic recess to enlarge (dotted arrows). R: inferior margin of the lung located within the costodiaphragmatic recess.

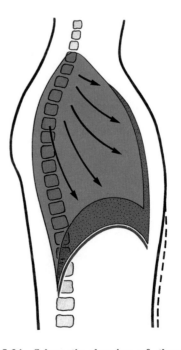

5.34 Schematic drawing of thoracic volume changes in ventilation (lateral view). Arrows indicate direction of distension of the lung.

6. Abdominal Organs

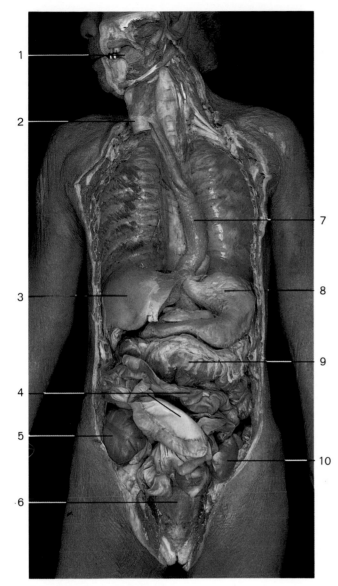

6.1   Digestive organs *in situ*.

6.2   Organization and location of abdominal organs (schematic drawing).

1	Tongue	6	Rectum	11	Oral cavity	16	Vermiform appendix
2	Trachea	7	Esophagus	12	Gallbladder	17	Oropharynx
3	Liver	8	Stomach	13	Duodenum	18	Pancreas
4	Small intestine	9	Transverse colon	14	Ascending colon	19	Descending colon
5	Cecum	10	Sigmoid colon	15	Ileum	20	Jejunum

The digestive system consists of the organs of the alimentary tract and glands for exocrine and endocrine secretion. The system may be summarized as follows:

Salivary glands   Liver, Pancreas
↓   ↓

Oral cavity → Pharynx → Esophagus → Stomach → Small intestine → Large intestine → Anus

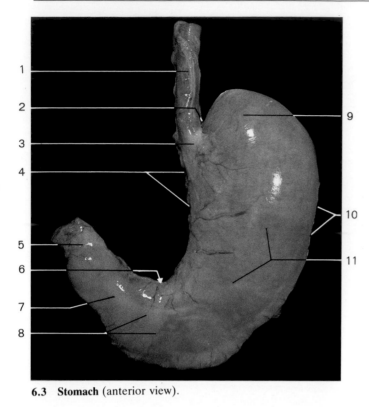

6.3　Stomach (anterior view).

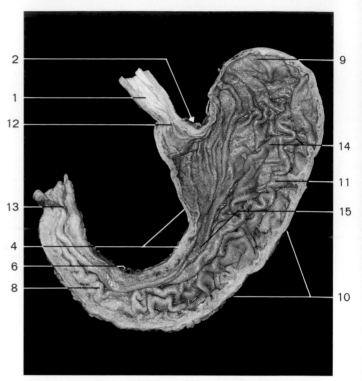

6.4　Mucosa of posterior wall of stomach (anterior view).

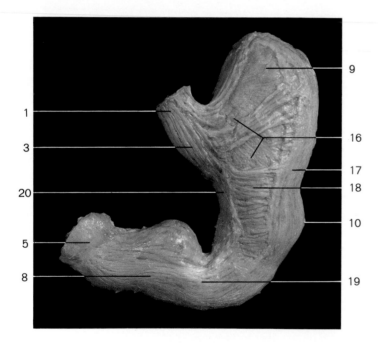

Folds (*rugae*) of the gastric mucosa are distended when the stomach is filled with ingested foods or liquids. The muscle coat of the stomach consists of three layers: longitudinal, circular and oblique. The oblique muscle is distinct only near the cardia.

6.5　Muscular coat of stomach, inner layer (anterior view). At the upper part of the stomach the circular muscle layer is opened to show the oblique muscle layer.

1　Esophagus	11　Body of stomach
2　Cardiac notch	12　Cardiac orifice
3　Cardiac part of stomach	13　Pyloric canal
4　Lesser curvature of stomach	14　Folds of mucous membrane (*rugae*)
5　Pyloric sphincter	15　Gastric canal
6　Angular notch	16　Oblique muscle fibers
7　Pyloric antrum	17　Longitudinal muscle layer of greater curvature
8　Pyloric part of stomach	18　Circular muscle of body of stomach
9　Fundus of stomach	19　Longitudinal muscle layer, transition from body to pyloric part
10　Greater curvature of stomach	20　Longitudinal muscle layer of lesser curvature

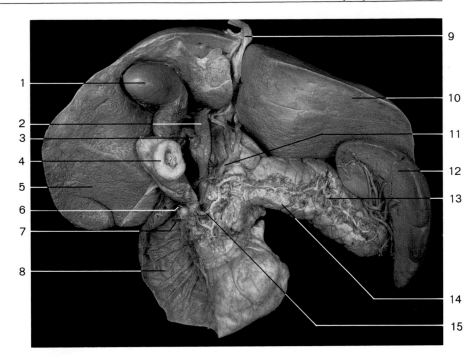

6.6 Pancreas with adjacent duodenum, spleen, gallbladder and liver (anterior view). The liver is slightly reflected upward. The duodenum has been opened and the pancreatic duct dissected. The pancreatic duct and accessory pancreatic duct are separate because of the independent development of the ventral and dorsal parts of the pancreas and their ducts.

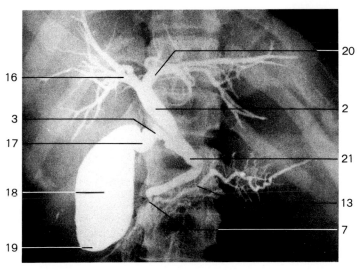

6.7 Postmortem cholangiopancreatogram.

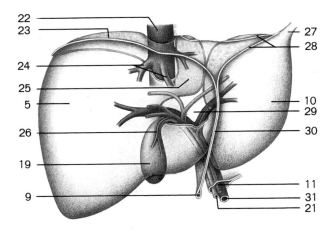

6.8 Schematic drawing of the liver with margins of peritoneal folds (anterior view).

1	Gallbladder	12	Spleen
2	Common hepatic duct	13	Pancreatic duct
3	Cystic duct	14	Pancreas
4	Pylorus (cut)	15	Accessory pancreatic duct
5	Right lobe of liver	16	Right hepatic duct
6	Lesser duodenal papilla (probe)	17	Neck of gallbladder
7	Greater duodenal papilla (probe)	18	Body of gallbladder
8	Duodenum	19	Fundus of gallbladder
9	Ligamentum teres	20	Left hepatic duct
10	Left lobe of liver	21	Common bile duct
11	Hepatic artery proper	22	Inferior vena cava

23	Peritoneum
24	Hepatic veins
25	Caudate lobe of liver
26	Cystic artery
27	Fibrous appendix of liver
28	Coronary ligament of liver
29	Porta hepatis
30	Falciform ligament of liver
31	Portal vein

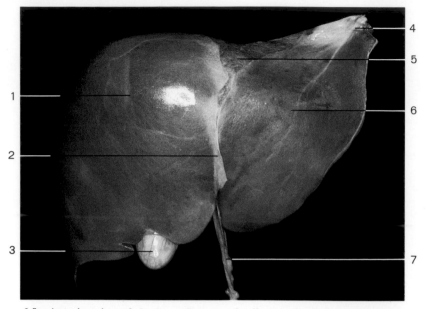

1	Right lobe
2	Falciform ligament
3	Gallbladder
4	Fibrous appendix
5	Bare area
6	Left lobe
7	Ligamentum teres
8	Cystic duct
9	Common bile duct
10	Inferior vena cava
11	Caudate lobe
12	Quadrate lobe
13	Hepatic artery
14	Portal vein
15	Ligamentum venosum

6.9   Anterior view of the liver. Because the liver is fixed to the diaphragm at the bare area, it moves with respiration and helps interhepatic circulation. The ligamentum teres is a remnant of the umbilical vein of the fetal circulation (see page 60).

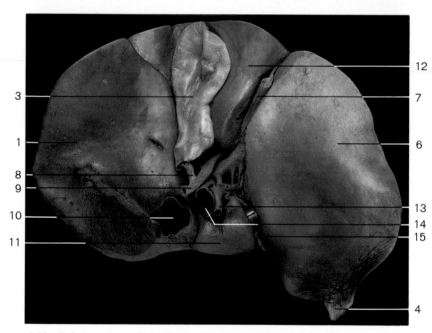

6.10   Inferior view of the liver. To demonstrate the ligamentum venosum a short rod is inserted between the caudate and left lobes. The ligamentum venosum and ligamentum teres are remnants of the fetal circulation.

The bile produced in the liver flows through bilateral hepatic ducts to empty into the common bile duct. This duct, along with the pancreatic ducts, opens into the duodenum. When the duodenum is devoid of food, bile enters the cystic duct, that opens into the midportion of the common bile duct, and is conveyed to the gallbladder where it is condensed and stored.

1 Liver
2 Middle colic artery
3 **Superior mesenteric artery**
4 Right colic artery
5 Ileocolic artery
6 Lymph nodes
7 Ileal artery
8 Cecum
9 Ileum
10 Greater omentum
11 Transverse colon
12 Jejunal artery
13 Jejunum
14 Gallbladder
15 Cystic vein
16 **Hepatic portal vein**
17 Duodenum
18 Pancreas
19 **Superior mesenteric vein**
20 Right gastroepiploic vein
21 Tributaries of superior mesenteric
   vein
22 Inferior vena cava
23 Stomach
24 Gastric veins
   a Coronary vein
   b Pyloric vein
25 **Splenic vein**
26 Short gastric veins
27 Spleen
28 Pancreatic veins
29 Left gastroepiploic vein
30 **Inferior mesenteric vein**
31 Tributaries of inferior mesenteric vein
32 Large intestine (sigmoid colon)

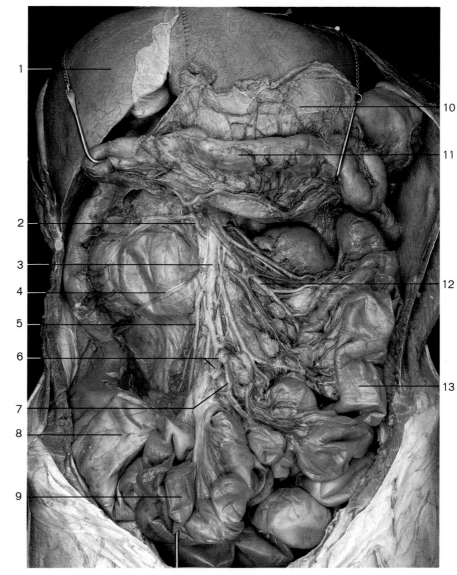

6.11 Distribution of the superior mesenteric artery which supplies the small intestine and proximal half of the large intestine. The greater omentum has been reflected upward.

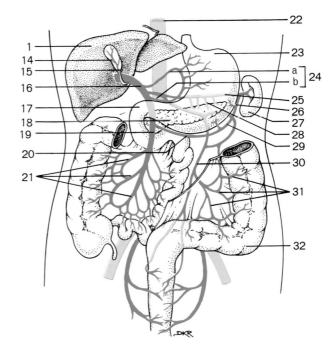

6.12 The tributaries of the hepatic portal vein (schematic drawing). (From Weinreb, E. L.: *Anatomy and Physiology*. Addison-Wesley Publishing Co., Reading, Mass., 1984.)

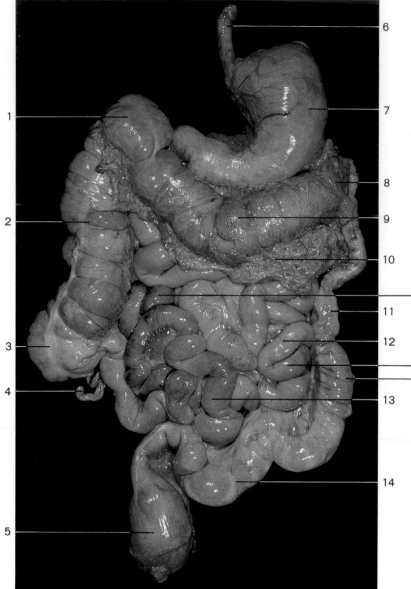

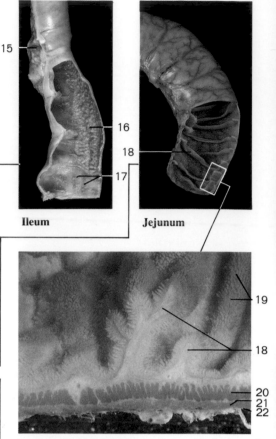

Ileum

Jejunum

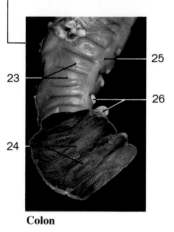

Mucosa of small intestine showing villi (enlarged).

Colon

6.13   Gastrointestinal tract. The portion extending from the lower esophagus to the anus has been removed from the peritoneal cavity. Anatomical features at four different levels of the intestine are shown in accompanying photographs. Circular folds become less prominent in the distal
part of the small intestine. The intestinal villi, prominent in the small intestine, are not present in the colon. Free teniae and epiploic appendages are
provided on the external surface of the colon.

1	Hepatic (right) flexure of colon	
2	Ascending colon	
3	Cecum	
4	Vermiform appendix	
5	Rectum	
6	Esophagus (cut)	
7	Stomach	
8	Splenic (left) flexure of colon	
9	Transverse colon	
10	Greater omentum	
11	Descending colon	
12	Jejunum	
13	Ileum	

14	Sigmoid colon
15	**Cut edge of mesentery**
16	Aggregated lymphatic nodules
17	Solitary lymphatic nodules
18	Circular folds
19	Intestinal villi
20	Circular muscle layer
21	Longitudinal muscle layer
22	Serosa
23	Haustra of colon
24	Semilunar folds
25	Free tenia
26	Epiploic appendages

6.14 and 6.15 ▷

1	Right and left lobes of liver
2	Ligamentum teres
3	Hepatic plexus
4	Hepatic artery proper and portal vein
5	Cystic artery
6	Gallbladder and common bile duct
7	Gastroduodenal artery
8	**Common hepatic artery**
9	**Splenic artery**
10	Descending part of duodenum
11	Right gastroepiploic artery
12	Colic flexure
13	Right colic artery
14	Superior mesenteric vein
15	Superior mesenteric plexus
16	Duodenum

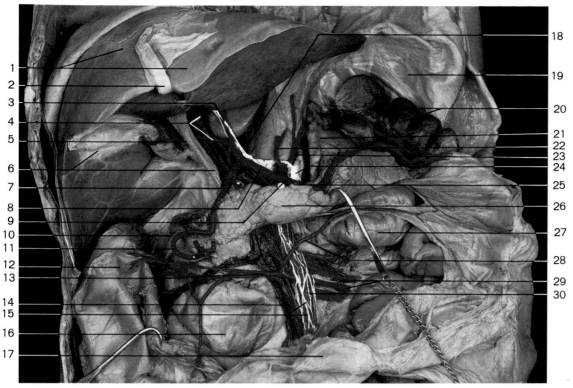

6.14   Vessels and autonomic nerves of the upper part of the abdominal cavity. The stomach has been removed and the liver and pancreas have been slightly reflected. Part of the transverse mesocolon and the parietal layer of the peritoneum have been removed to show the vessels and nerves on the posterior abdominal wall.

Green=bile ducts;   red=arteries;   blue=veins;   white=autonomic nerves and plexus.

17	Transverse colon
18	Inferior vena cava
19	Diaphragm
20	Spleen
21	Adrenal (suprarenal) gland
22	Left gastric artery
23	Inferior phrenic artery
24	Celiac ganglion
25	**Celiac trunk**
26	Pancreas
27	Jejunum
28	Middle colic artery
29	Jejunal arteries
30	**Superior mesenteric artery**
31	Caudate lobe of liver
32	Right gastric artery
33	Ileocolic artery
34	Terminal part of ileum (cut)
35	Vermiform appendix
36	Right common iliac artery
37	Cecum
38	Cardioesophageal branch of left gastric artery
39	Left gastroepiploic artery
40	Stomach
41	Renal artery
42	Left testicular artery
43	Left kidney
44	Left colic artery
45	Sigmoid colon

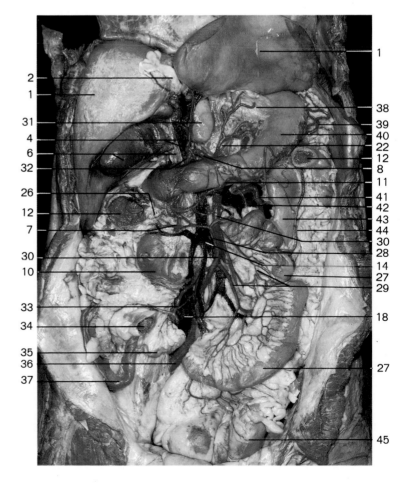

6.15   Vessels of abdominal organs (anterior view). The transverse colon and parts of the small intestine and lesser omentum have been removed. The liver has been raised. Red=arteries;   blue=veins.

7. Urogenital and Retroperitoneal Organs

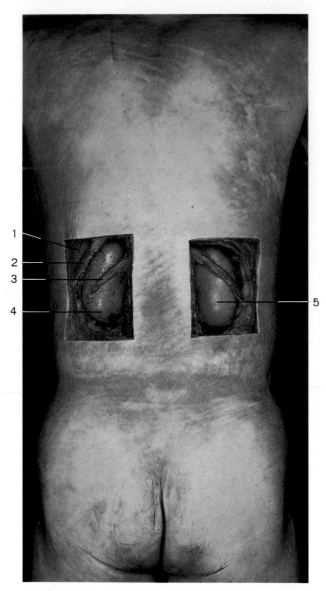

7.1 Kidneys *in situ*. The back is opened to show the retroperitoneal location of the kidneys. The right kidney is slightly more inferior than the left kidney due to the presence of the liver.

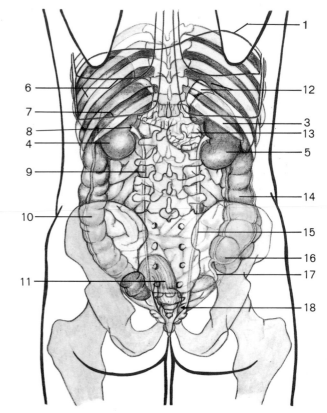

7.2 Location of urinary organs (posterior view). The superior (cranial) part of the kidney reaches the level of the 11th rib and margin of pleura and lung; the inferior (caudal) part reaches the level of the 3rd lumbar vertebra.

1	Diaphragm	7	Margin of pleura
2	Eleventh rib	8	Renal pelvis
3	Twelfth rib	9	Left ureter
4	Left kidney	10	Descending colon
5	Right kidney	11	Rectum
6	Spleen and margin of lung	12	Liver and right adrenal (suprarenal) gland
13	Pancreas		
14	Ascending colon		
15	Right ureter		
16	Cecum		
17	Vermiform appendix		
18	Position of urinary bladder		

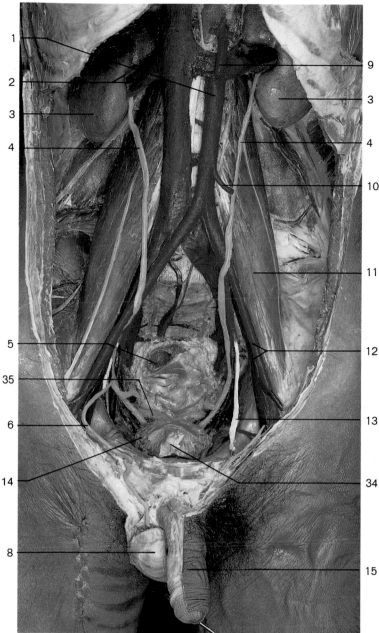

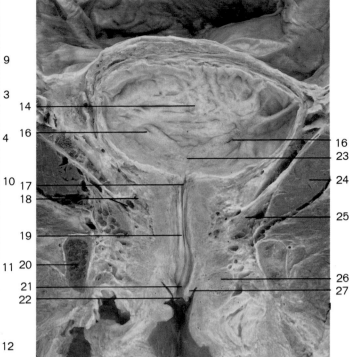

7.4 Frontal (coronal) section through the female urinary bladder and urethra (anterior view).

7.3 Male urogenital system *in situ* (anterior view). The peritoneum has been removed. The umbilical artery is obliterated after birth and remains as the umbilical ligament.
Red = arteries; blue = veins; orange = ureter; yellow = umbilical ligament; green = ductus deferens

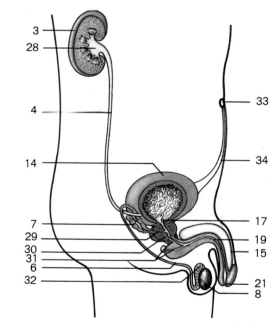

7.5 Midsagittal section through the male urogenital system (schematic drawing).

1 Aorta	13 Medial umbilical ligament (containing remnants of umbilical artery)	24 Obturator internus
2 Renal artery and vein		25 Levator ani
3 Kidney	14 Urinary bladder	26 Bulb of vestibule
4 Ureter	15 Penis	27 Left labium minus
5 Rectum (cut)	16 Ureteric orifice	28 Renal pelvis
6 Ductus deferens	17 Internal urethral orifice	29 Ejaculatory duct
7 Seminal vesicle	18 Vesicouterine venous plexus	30 Prostate gland
8 Testis	19 Urethra	31 Bulbourethral or Cowper's gland and duct
9 Superior mesenteric artery	20 Pubic bone (cut)	32 Epididymis
10 Inferior mesenteric artery	21 External urethral orifice	33 Umbilicus
11 Psoas muscle	22 Vestibule of vagina	34 Median umbilical ligament
12 External iliac artery and vein	23 Trigone of urinary bladder	35 Ampulla of ductus deferens

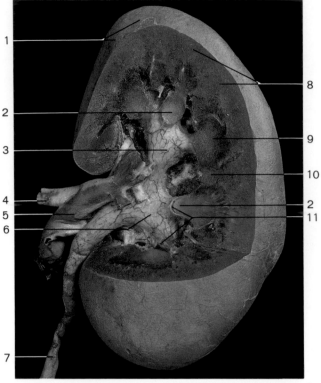

1
2
3
4
5
6
7
8
9
10
2
11

7.6   Right kidney (posterior view). The vessels are injected with colored solutions.

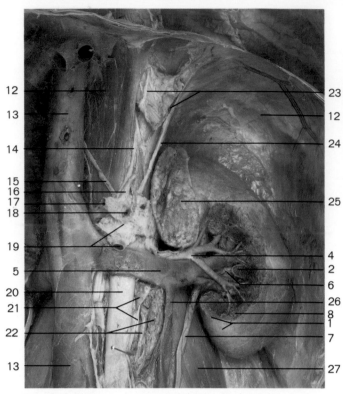

12
13
14
15
16
17
18
19
5
20
21
22
13
23
12
24
25
4
2
6
26
8
1
7
27

7.7   Left kidney and adrenal gland *in situ* (anterior view). The anterior part of the kidney has been removed to show the vessels of the kidney.

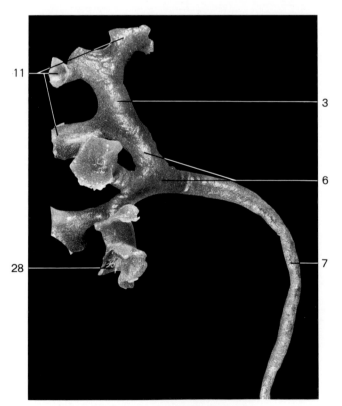

11
28
3
6
7

7.8   Cast of renal pelvis and proximal ureter.

1   Fibrous capsule of kidney
2   Renal papilla (cut surface)
3   Major calyx
4   Renal artery
5   Renal vein
6   Renal pelvis
7   Ureter
8   Cortex of kidney
9   Medulla of kidney
10   Renal columns
11   Minor calyces
12   Diaphragm
13   Inferior vena cava
14   Right vagus nerve
15   Right inferior phrenic artery
16   Left gastric artery
17   Common hepatic artery (cut)
18   Splenic artery (cut)
19   Superior mesenteric artery (cut) and celiac ganglion
20   Abdominal aorta
21   Sympathetic trunk with ganglion
22   Lymph node and vessels
23   Esophagus (abdominal part) and left vagus nerve
24   Left inferior phrenic artery
25   Left adrenal gland
26   Left testicular vein
27   Psoas major
28   Papillary ducts

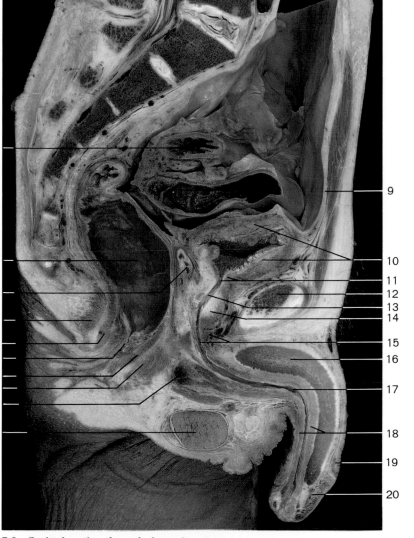

1   Sigmoid colon
2   Ampulla of rectum
3   Ampulla of ductus deferens
4   Sphincter ani externus
5   Sphincter ani internus
6   Anal canal
7   Bulb of penis
8   Testis
9   Median umbilical ligament
10 Urinary bladder
11 Internal urethral orifice and sphincter
    (sphincter vesicae)
12 Pubic symphysis
13 Prostatic urethra
14 Prostate gland
15 Membranous urethra and external urethral
    sphincter (sphincter urethrae)
16 Corpus cavernosum of penis
17 Penile or spongy urethra
18 Corpus spongiosum of penis
19 Foreskin or prepuce
20 Glans penis
21 Ureters
22 Seminal vesicles
23 Urogenital diaphragm and membranous
    urethra
24 Bulbourethral or Cowper's gland
25 Left and right crus penis
26 Epididymis
27 Ductus deferens

7.9   Sagittal section through the male pelvis.

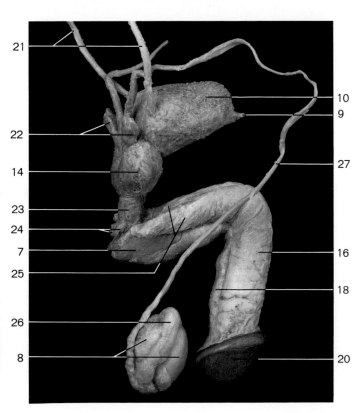

7.10   Male reproductive organs, isolated
(right lateral view).

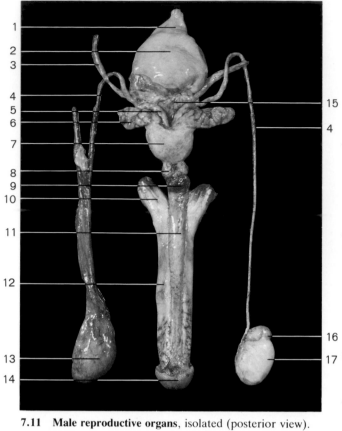

7.11   Male reproductive organs, isolated (posterior view).

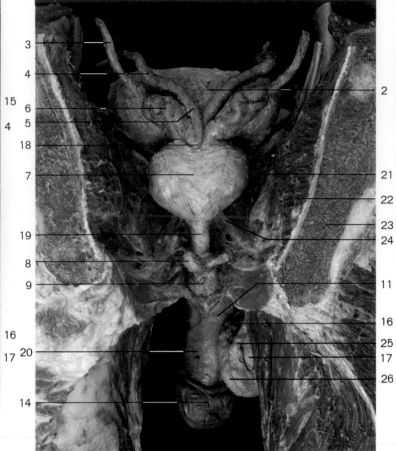

7.12   Accessory glands of male reproductive organs *in situ.*
Frontal (coronal) section through the pelvic cavity. Posterior
view of urinary bladder, prostate and seminal vesicles.

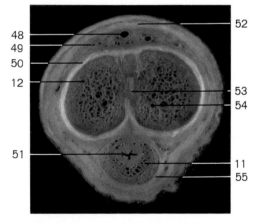

**7.14   Cross section through penis showing
erectal tissues.**

7.13   Pelvic cavity in the male with vessels and nerves (medial
view of left half). The urinary bladder has been reflected and
the parietal layer of peritoneum has been removed.
Red = arteries; blue = veins; orange = ureter; green = ductus
deferens and seminal vesicle; white = nerves; yellow =
medial umbilical ligament.

1 Median umbilical ligament
2 Urinary bladder
3 Ureter
4 Ductus deferens
5 Ampulla of ductus deferens
6 Seminal vesicle
7 Prostate gland
8 Bulbourethral or Cowper's glands
9 Bulb of penis
10 Crus penis
11 Corpus spongiosum of penis
12 Corpus cavernosum of penis
13 Testis and epididymis with coverings
14 Glans penis
15 Fundus of urinary bladder
16 Head of epididymis
17 Testis
18 Ejaculatory duct
19 Membranous urethra
20 Penis
21 Levator ani
22 Obturator internus
23 Pelvic bone (cut)
24 Puboprostatic ligament
25 Beginning of ductus deferens
26 Tail of epididymis
27 Right common iliac artery and vein
28 Median sacral artery
29 Right external iliac artery (cut)
30 Right internal iliac artery (cut)
31 Superior rectal artery
32 Inferior gluteal artery
33 Internal pudendal artery
34 Rectum
35 Vesicoprostatic venous plexus
36 Left common iliac artery
37 Iliacus muscle
38 Lateral cutaneous femoral nerve
39 Femoral nerve
40 Left internal iliac artery
41 Left external iliac artery and vein
42 Inferior epigastric artery and vein
43 Umbilical artery
44 Medial umbilical ligament
45 Superior vesical artery
46 Obturator vein, artery, and nerve
47 Pubic bone (cut)
48 Deep dorsal vein of penis
49 Dorsal artery of penis
50 Tunica albuginea of corpus cavernosum
51 Urethra
52 Deep fascia of penis
53 Septum pectiniforme
54 Deep artery of penis
55 Tunica albuginea of corpus spongiosum
56 Right common iliac artery
57 Iliopsoas muscle
58 Right external iliac vein
59 Inferior epigastric artery and vein
60 Pubic symphysis
61 Root of penis
62 Iliolumbar artery
63 Lateral sacral artery
64 Superior gluteal artery
65 Pudendal and coccygeal plexuses
66 Coccygeus muscle
67 Middle rectal vein, inferior vesical vein
68 Middle rectal artery
69 Superior vesical artery and branch to the ductus deferens
70 Pudendal nerve
71 Membranous and penile urethra
72 Urogenital diaphragm
73 Obturator artery
74 Inferior vesical artery
75 Inferior rectal artery

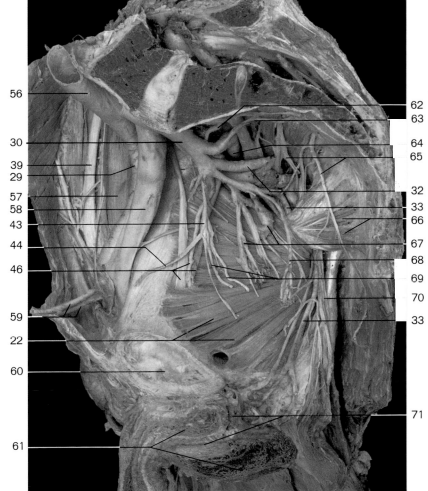

7.15 Vessels of pelvic cavity in the male (sagittal section, right side, medial view).

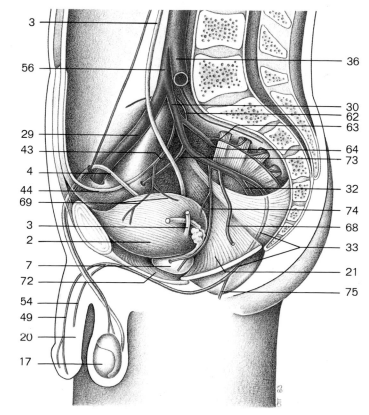

7.16 Main branches of internal iliac artery in the male (schematic drawing).

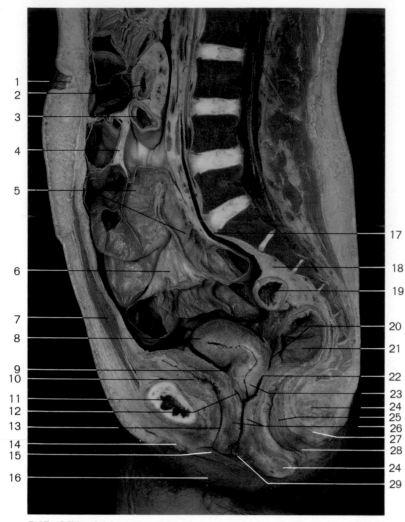

1	Umbilicus
2	Duodenum
3	Ascending part of duodenum
4	Root of mesentery
5	Small intestine
6	Mesentery
7	Rectus abdominis
8	Fundus of uterus
9	Vesicouterine pouch
10	Urinary bladder
11	Pubic symphysis
12	Anterior fornix of vagina
13	Urethra
14	Clitoris '
15	Labium minus
16	Labium majus
17	Promontory
18	Mesosigmoid
19	Sigmoid colon
20	Rectouterine pouch
21	Rectum
22	Posterior fornix of vagina
23	Cervix
24	Sphincter ani externus
25	Anal canal
26	Vagina
27	Sphincter ani internus
28	Anus
29	Hymen
30	Infundibulum of uterine tube
31	Ampulla of uterine tube
32	Ovary
33	Deep transversus perinei muscle
34	Kidney
35	Abdominal part of ureter
36	Pelvic part of ureter
37	Perineum

7.17 Midsagittal section through the female trunk. The urinary bladder is empty, the position and shape of the uterus are normal.

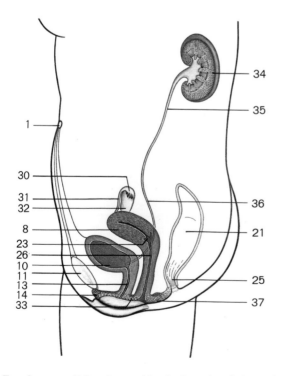

7.18   Female urogenital system, midsagittal section (schematic drawing).

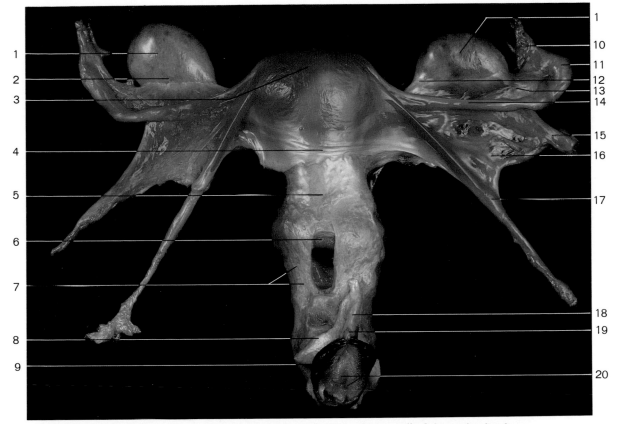

7.19 Female reproductive organs, isolated (anterior view). The anterior wall of the vagina has been opened to show the vaginal part of the cervix.

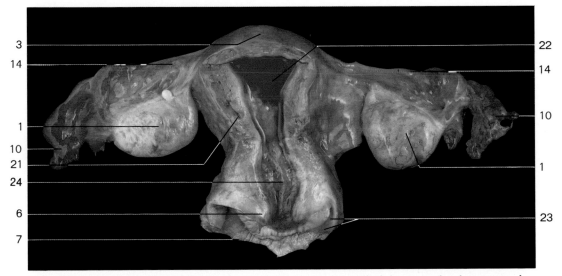

7.20 Uterus and related organs (posterior view). The posterior wall of the uterus has been opened.

1 Ovary	13 Mesosalpinx
2 Mesovarium	14 Isthmus of uterine tube
3 Fundus of uterus	15 Suspensory ligament of ovary
4 Vesicouterine pouch	(caudally displaced)
5 Cervix of uterus	16 Broad ligament of uterus
6 Vaginal part of cervix	17 Round ligament of uterus
7 Vagina	18 Corpus cavernosum of clitoris
8 Crus of clitoris	19 Glans of clitoris
9 Labium minus	20 External urethral orifice, hymen
10 Fimbriae of uterine tube	21 Body of uterus
11 Infundibulum of uterine tube	22 Mucous membrane of uterus (hyperemic)
12 Ligament of ovary	23 Anterior fornix of vagina
	24 Cervical canal

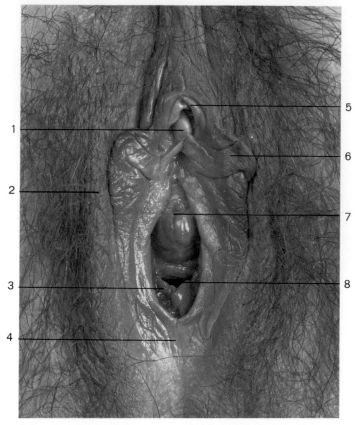

7.21   External genital organs in the virgin female (anterior view).

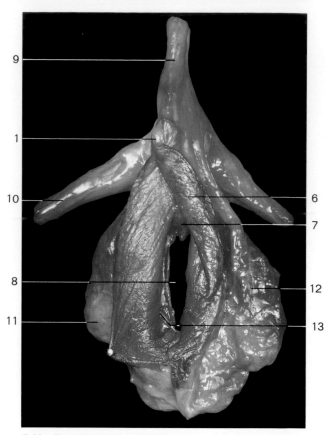

7.22   Female external genital organs, isolated to show the glands and bulbs of vestibule.

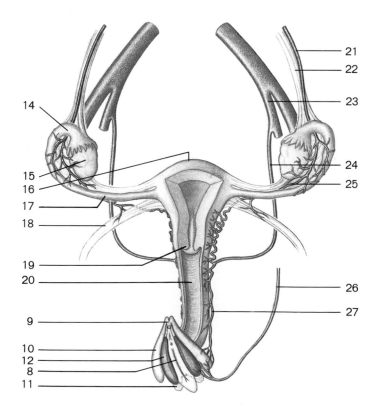

7.23   Arteries of female reproductive organs (schematic drawing).

1   Glans of clitoris
2   Labium majus
3   Hymen
4   Posterior labial commissure
5   Prepuce of clitoris
6   Labium minus
7   External orifice of urethra
8   Vaginal orifice
9   Body of clitoris
10   Crus of clitoris
11   Greater vestibular gland
12   Bulb of vestibule
13   Opening of greater vestibular gland (probe)
14   Infundibulum of uterine tube
15   Ovary
16   Fundus of uterus
17   Isthmus of uterine tube
18   Round ligament of uterus
19   Vaginal part of cervix
20   Vagina
21   Ovarian artery
22   Suspensory ligament of ovary
23   Internal iliac artery
24   Uterine artery
25   Ovarian branch of uterine artery (anastomoses with ovarian artery)
26   Internal pudendal artery
27   Vaginal artery

8. Head and Neck

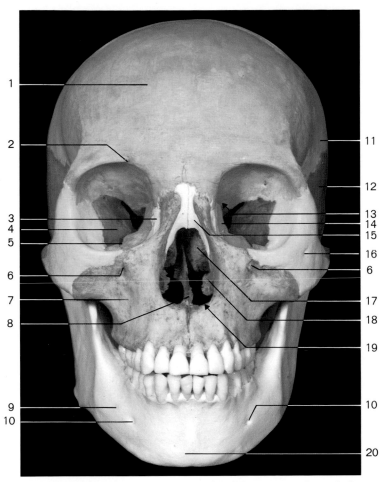

1 Frontal bone (light brown)
2 Supraorbital notch
3 Lacrimal bone (yellow)
4 Greater wing of sphenoid bone (red)
5 Inferior orbital fissure
6 Infraorbital foramen
7 Maxilla (violet)
8 Vomer (orange)
9 Mandible (white)
10 Mental foramen
11 Parietal bone (light green)
12 Squamous part of temporal bone (brown)
13 Optic foramen or canal
14 Ethmoid bone (dark green)
15 Nasal bone (white)
16 Zygomatic bone (yellow)
17 Middle nasal concha
18 Inferior nasal concha (pink)
19 Anterior nasal aperture (piriform aperture)
20 Mental protuberance
21 Anterior fontanel
22 Coronal suture
23 Frontal suture
24 Sagittal suture
25 Posterior fontanel
26 Occipital bone (blue)
27 Lambdoidal suture

8.1 Skull of the adult, anterior view (skull bones are color coded).

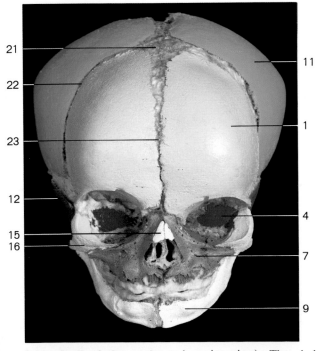

8.2A Skull of the newborn (anterior view). The skull bones of the neonate are loosely jointed to facilitate passage through the birth canal during delivery.

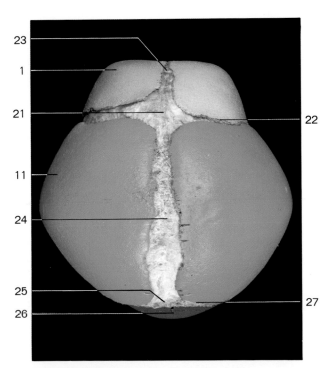

8.2B Skull of the newborn (superior view). Facial part on top.

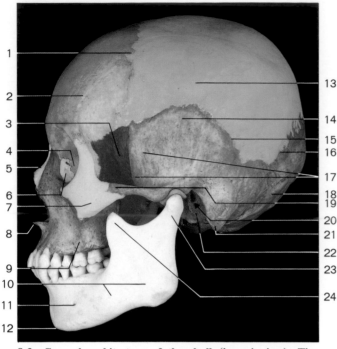

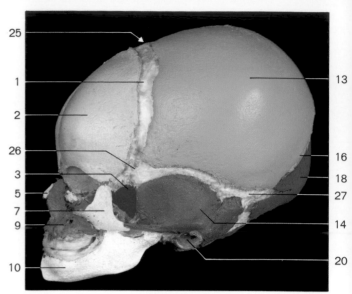

8.4   Skull of the newborn (lateral view).

8.3   General architecture of the skull (lateral view). The base of the skull is the border between the cranial and facial parts of the skull. The cranial bones protect the brain and sense organs (for information processing, sight, hearing and balance). The facial bones support the nasal and oral cavities and hold the teeth (for respiration and digestion), as well as protect the organs for sight and smell.

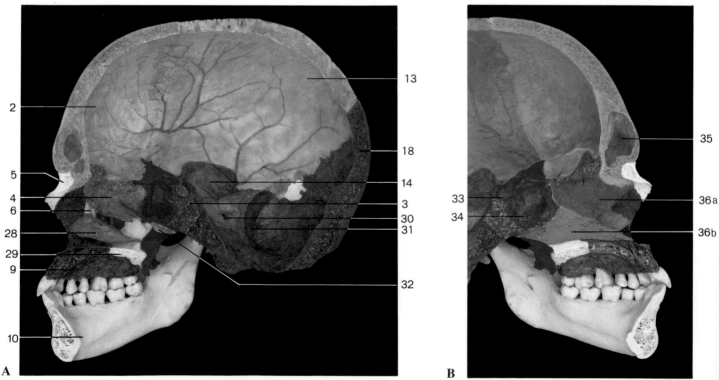

A   B

8.5   Parasagittal section through the skull. Internal view of the right half (A) does not include the nasal septum. View of the left half (B) includes the nasal septum.

1 Coronal suture
2 Frontal bone (light brown)
3 Sphenoid bone (red)
4 Ethmoid bone (dark green)
5 Nasal bone (white)
6 Lacrimal bone (yellow)
7 Zygomatic bone (yellow)
8 Anterior nasal spine
9 Maxilla (violet)
10 Mandible (white)
11 Mental foramen
12 Mental protuberance
13 Parietal bone (light green)
14 Squamous part of temporal bone (brown)
15 Squamosal suture
16 Lambdoidal suture
17 Temporal fossa
18 Occipital bone (blue)
19 Zygomatic arch
20 External acoustic meatus
21 Mastoid process
22 Tympanic portion of temporal bone (dark brown)
23 Condylar process of mandible
24 Coronoid process of mandible
25 Anterior fontanel
26 Anterolateral (sphenoidal) fontanel
27 Posterolateral (mastoid) fontanel
28 Inferior nasal concha (pink)
29 Palatine bone (white)
30 Internal acoustic meatus
31 Groove for sigmoid sinus
32 Ala of vomer (orange)
33 Hypophyseal fossa of sella turcica
34 Sphenoidal sinus
35 Frontal sinus
36 Osseous part of nasal septum
 a Perpendicular plate (dark green)
 b Vomer (orange)
37 Pterygoid canal (nerve of pterygoid canal)
38 Foramen ovale (mandibular nerve), spinous foramen (middle meningeal artery)
39 Internal carotid artery (carotid canal), internal jugular vein (venous part of jugular foramen)
40 Stylomastoid foramen (facial nerve)
41 Jugular foramen (glossopharyngeal, vagus and accessory nerves)
42 Hypoglossal canal (hypoglossal nerve)
43 Incisive bone (dark violet)
44 Oculomotor and trochlear nerves (superior orbital fissure)
45 Ophthalmic nerve (superior orbital fissure)
46 Maxillary nerve (foramen rotundum)
47 Mandibular nerve (foramen ovale)
48 Middle meningeal artery (spinous foramen)
49 Internal carotid artery (carotid canal)
50 Vestibulocochlear and facial nerves (internal acoustic meatus)
51 Jugular vein (jugular foramen)
52 Optic nerve (optic canal)
53 Foramen rotundum (maxillary nerve)
54 Spinous foramen (middle meningeal artery)
55 Foramen lacerum
56 Abducens nerve (superior orbital fissure)

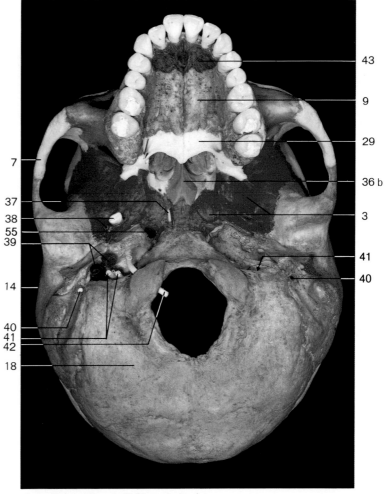

8.6 Base of the skull (from below).

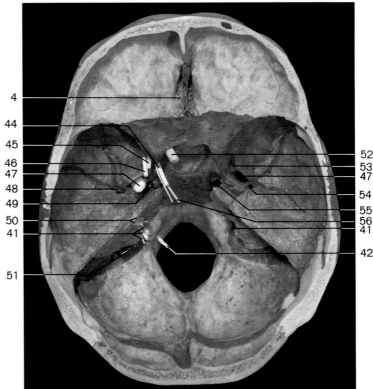

8.7 Base of the skull (from above). On the left side rods have been inserted into the foramina or canals to show the locations of the nerves or blood vessels.

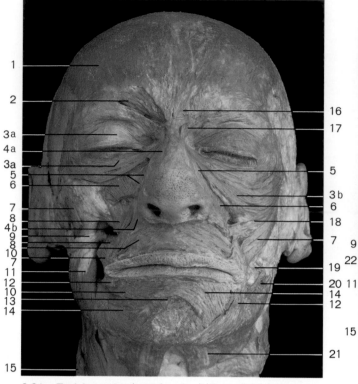

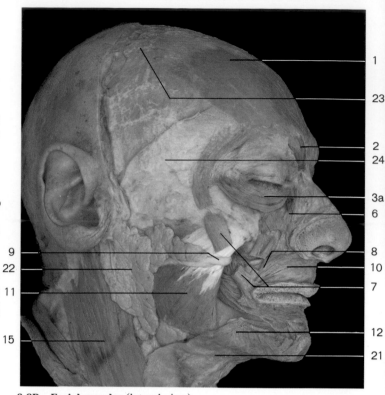

8.8A   Facial muscles (anterior view). Superficial layer (left side); deeper layer (right side).

8.8B   Facial muscles (lateral view).

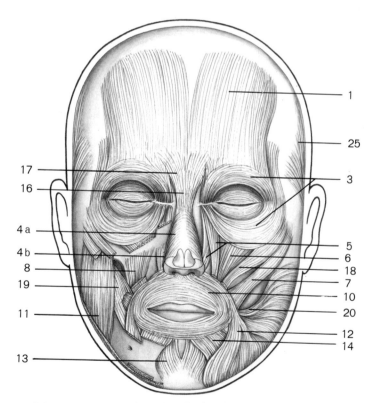

8.9   Facial muscles (anterior view), schematic drawing.

1 Frontal belly of occipitofrontalis
2 Corrugator supercilii
3 Orbicularis oculi
   a   Palpebral part
   b   Orbital part
4 Nasalis
   a   Transverse part
   b   Alar part
5 Levator labii superioris alaeque nasi
6 Levator labii superioris
7 Zygomaticus major
8 Levator anguli oris
9 Parotid duct
10 Orbicularis oris
11 Masseter
12 Depressor anguli oris
13 Mentalis
14 Depressor labii inferioris
15 Sternocleidomastoid
16 Procerus
17 Depressor supercilii
18 Zygomaticus minor
19 Buccinator
20 Risorius
21 Platysma
22 Parotid gland
23 Galea aponeurotica
24 Temporal fascia
25 Temporoparietalis (auricularis superior)

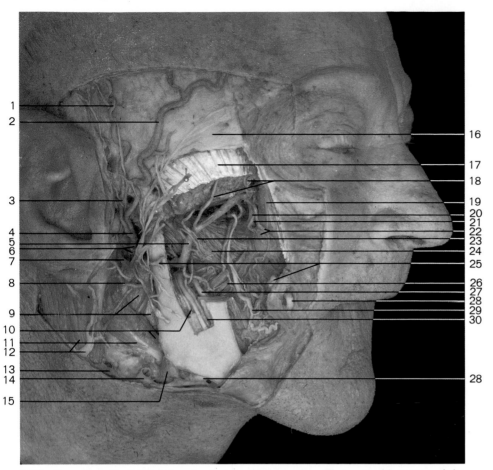

1 Parietal branch of superficial temporal artery
2 Frontal branch of superficial temporal artery
3 Auriculotemporal nerve
4 Superficial temporal artery
5 Maxillary artery
6 Communicating branches between facial and auriculo-temporal nerves
7 Facial nerve
8 Posterior auricular artery
9 Internal jugular vein
10 Mylohyoid nerve, stylohyoid muscle
11 Posterior belly of digastric muscle
12 Great auricular nerve, sterno-cleidomastoid
13 External jugular vein
14 Retromandibular vein
15 Submandibular gland
16 Temporal fascia
17 Temporalis
18 Deep temporal arteries
19 Posterior superior alveolar nerve
20 Infraorbital artery
21 Sphenopalatine artery
22 Posterior superior alveolar artery
23 Masseteric artery and nerve
24 Lateral pterygoid
25 Transverse facial artery, parotid duct
26 Medial pterygoid
27 Lingual nerve
28 Facial artery
29 Buccal nerve and artery
30 Inferior alveolar artery and nerve (mandibular canal opened)
31 Inferior alveolar nerve
32 Superior constrictor muscle of pharynx
33 Styloglossus
34 Submandibular ganglion
35 Hypoglossal nerve
36 Greater horn of hyoid bone
37 Stylohyoid
38 Inferior constrictor muscle of pharynx
39 Esophagus
40 Trachea
41 Buccinator, parotid duct
42 Tongue
43 Genioglossus
44 Geniohyoid
45 Mylohyoid (reflected)
46 Hyoglossus
47 Thyroid cartilage of larynx

8.10 Deep dissection of facial and retromandibular region. The coronoid process of the mandible together with the insertions of the temporalis muscle have been removed to show the maxillary artery. The mandibular canal has been partly opened.

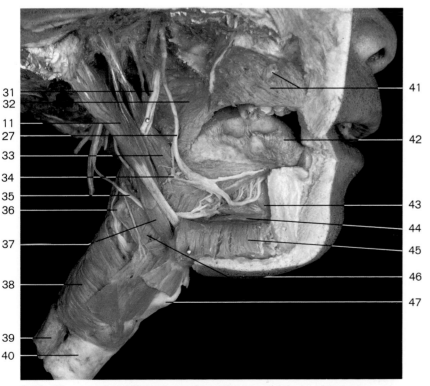

8.11 Pharyngeal and lingual muscles and nerves (right lateral view).

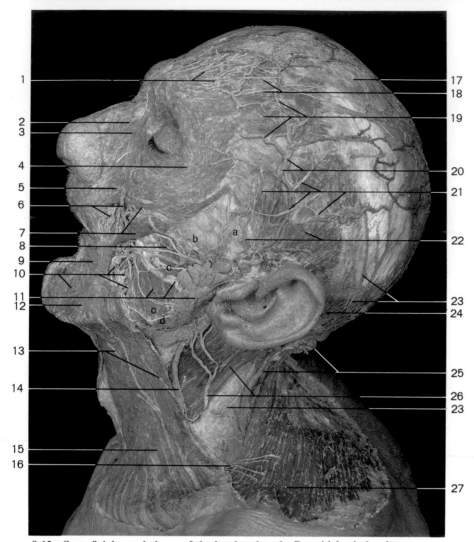

8.12   Superficial muscle layer of the head and neck. Parotid fascia has been removed. Branches of facial nerve: a=temporal branch; b=zygomatic branch; c=buccal branch; d=marginal mandibular branch.

8.13   Superficial nerves and vessels of the head (scheamtic drawing).

1   Medial branch of supraorbital nerve
2   Nasalis
3   Levator labii superioris alaeque nasi
4   Orbicularis oculi
5   Levator labii superioris
6   Facial artery and vein
7   Zygomaticus major and minor
8   Transverse facial artery
9   Orbicularis oris
10 Buccal nerves, depressor labii inferioris, facial artery and vein
11 Parotid gland and duct, masseter
12 Depressor anguli oris
13 Transverse cervical nerve
14 External jugular vein
15 Platysma
16 Supraclavicular nerve
17 Galea aponeurotica
18 Lateral branches of supra-orbital nerve
19 Frontal belly of occipitofrontalis, branches of superficial temporal vein and artery
20 Superficial temporal artery and vein
21 Auriculotemporal nerve
22 Zygomaticoorbital artery, temporoparietalis muscle
23 Lesser occipital nerve
24 Occipital belly of occipito-frontalis
25 Occipital artery and vein
26 Great auricular nerve, sterno-cleidomastoid
27 Trapezius
28 Angular artery
29 Branches of facial nerve
30 Parotid duct
31 Greater occipital nerve, occipital belly of epicranial muscle
32 Facial nerve
33 Superficial temporal artery, great auricular nerve
34 Parotid gland
35 Sternocleidomastoid, external carotid artery
36 Cervical branch of facial nerve

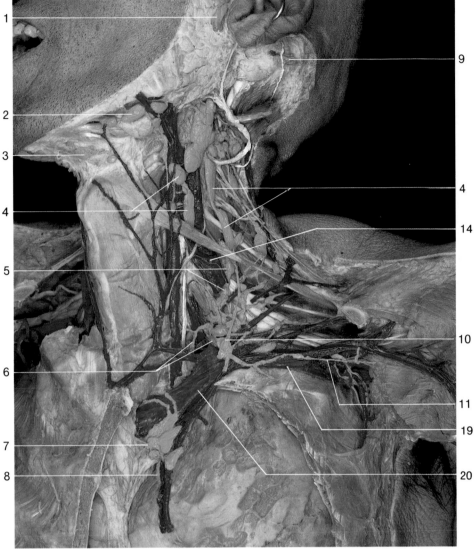

1 Parotid lymph nodes
2 Submandibular lymph nodes
3 Submental lymph nodes
4 Superior deep cervical lymph
   nodes, jugular trunk
5 Inferior deep cervical lymph
   nodes
6 Thoracic duct
7 Superior mediastinal lymph
   nodes
8 Bronchomediastinal trunk
9 Retroauricular lymph nodes
10 Supraclavicular lymph nodes
11 Left subclavian trunk
12 Facial vein
13 Occipital lymph nodes
14 Internal jugular vein
15 Deep cervical lymph nodes
   a Jugulodigastric lymph nodes
   b Juguloomohyoid lymph nodes
16 External jugular vein
17 Jugular trunk
18 Infraclavicular lymph nodes
19 Subclavian vein
20 Left brachiocephalic vein

8.14 Lymph nodes and lymphatic drainge of left side of neck. Sternocleidomastoid and clavicle
have been removed and left thorax opened. Lower part of the left internal jugular vein has been
cut and lifted to show the supraclavicular lymph nodes.

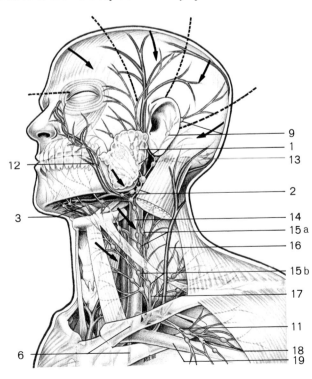

8.15 Lymph nodes and veins of the head and neck.
Dotted lines = border between irrigation areas;
arrows = direction of lymph flow.

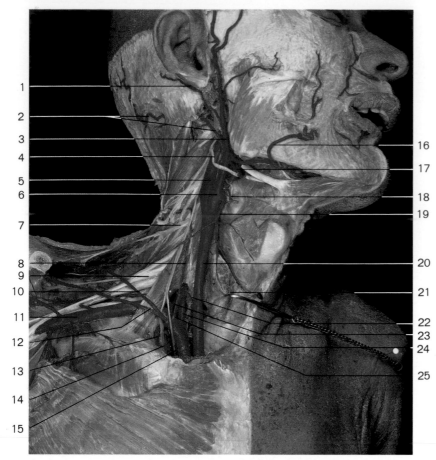

8.16 Arteries of right side of neck. Superficial dissection.

8.17 Arteries of right side of neck. Deep dissection. Maxillary artery (26) has been cut to show the internal carotid artery (5). Yellow = sympathetic trunk, green = lymphatic trunks.

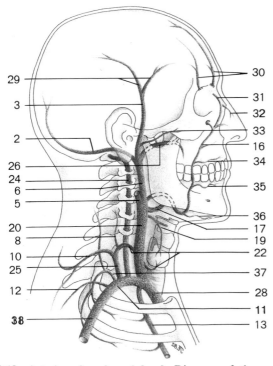

8.18 Arteries of neck and head. Diagram of the main branches of the external carotid (6) and subclavian (11) arteries.

1   Posterior auricular artery
2   Occipital artery
3   Superficial temporal artery
4   Hypoglossal nerve
5   **Internal carotid artery**
6   **External carotid artery**
7   Cervical plexus
8   Ascending cervical artery
9   Brachial plexus
10 Superficial cervical artery
11 **Subclavian artery**
12 Suprascapular artery
13 Internal thoracic artery
14 Vagus nerve
15 Phrenic nerve
16 Facial artery
17 Lingual artery
18 Superior laryngeal artery
19 Superior thyroid artery
20 **Common carotid artery**
21 Thyroid gland
22 Inferior thyroid artery
23 Deep cervical artery
24 Vertebral artery
25 Thyrocervical trunk
26 Maxillary artery
27 Internal jugular vein
28 **Brachiocephalic trunk**
29 Anterior and posterior branches of superficial temporal artery
30 Supraorbital and supratrochlear arteries
31 Angular artery
32 Dorsal nasal artery
33 Transverse facial artery
34 Superior labial artery
35 Inferior labial artery
36 Submental artery
37 Costocervical trunk
38 Axillary artery

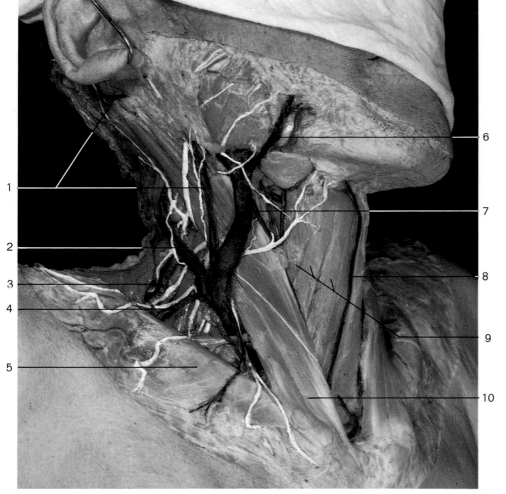

1   Posterior auricular vein
2   Occipital vein
3   Transverse cervical vein
4   Suprascapular vein
5   Clavicle
6   Facial vein
7   External jugular vein
8   Anterior jugular vein
9   Infrahyoid muscles
10 Sternocleidomastoid
11 Superficial temporal vein
12 Retromandibular vein
13 Trapezius
14 Internal jugular vein
15 Subclavian vein
16 Supraorbital vein
17 Middle temporal vein
18 Angular vein
19 Superior labial vein
20 Inferior labial vein
21 Submental vein
22 Superior thyroid vein
23 Jugular venous arch
24 Pterygoid plexus

8.19   Superficial veins of the neck, right side (lateral view).

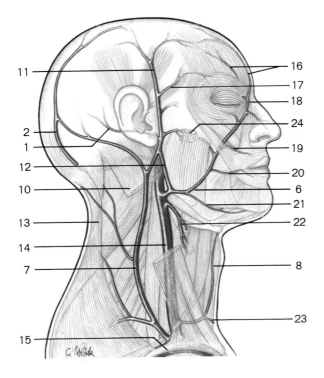

8.20   Veins of head and neck (schematic drawing).
Sternocleidomastoid has been partly removed to expose
the internal jugular vein. The pterygoid plexus (24) is connected posteriorly to the retromandibular vein by the
maxillary vein and connected anteriorly to the facial vein
by the deep facial vein.

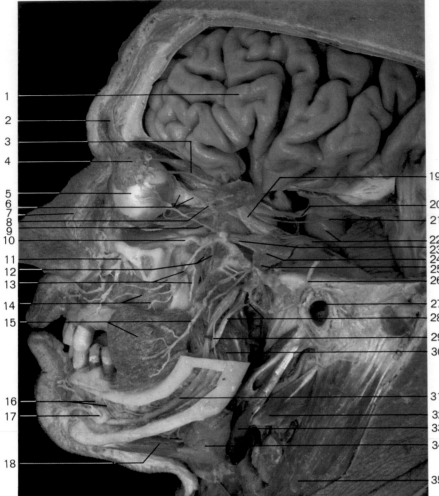

8.21 Dissection of the trigeminal nerve (V). The lateral wall of the cranial cavity, orbit, zygomatic arch and ramus of the mandible have been removed and the mandibular canal has been opened to show the nerve branches.

1   Cerebrum
2   Supraorbital nerve
3   Lacrimal nerve
4   Lacrimal gland
5   Eyeball
6   **Optic nerve** (II), short ciliary nerves
7   External nasal branch of anterior ethmoidal nerve
8   Ciliary ganglion
9   Zygomatic nerve
10 Infraorbital nerve
11 Terminal branches of infraorbital nerve
12 Pterygopalatine ganglion and nerves
13 Posterior superior alveolar branches
14 Superior dental plexus
15 Buccinator muscle, buccal nerve
16 Inferior dental plexus
17 Mental foramen, mental nerve
18 Anterior belly of digastric muscle
19 **Ophthalmic nerve** (branch of trigeminal)
20 **Oculomotor nerve** (III)
21 **Trochlear nerve** (IV)
22 **Trigeminal nerve** (V), pons
23 **Maxillary nerve** (branch of trigeminal)
24 Trigeminal ganglion
25 **Mandibular nerve** (branch of trigeminal)
26 Auriculotemporal nerve
27 External acoustic meatus (cut)
28 Lingual nerve, chorda tympani
29 Mylohyoid nerve
30 Medial pterygoid muscle
31 Inferior alveolar nerve
32 Posterior belly of digastric muscle
33 Stylohyoid
34 Submandibular gland
35 Sternocleidomastoid
36 Frontal nerve
37 Lacrimal nerve
38 Pterygopalatine ganglion
39 Location of otic ganglion
40 **Optic nerve** (II)
41 **Trigeminal nerve** (V)

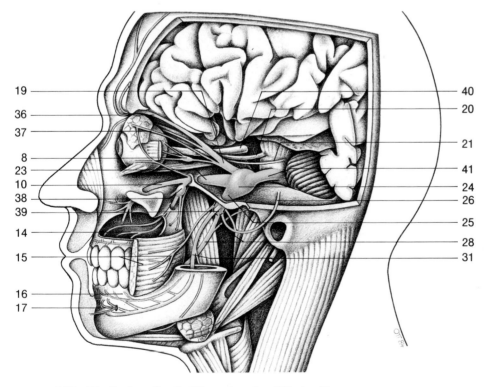

8.22 Distribution of optic (II), oculomotor (III), trochlear (IV) and trigeminal (V) cranial nerves.

1 Falx cerebri
2 Occipital lobe of cerebrum
3 Straight sinus
4 Tentorium cerebelli
5 Transverse sinus
6 Inferior colliculus of midbrain
7 Rhomboid fossa
8 Medulla oblongata
9 Posterior belly of digastric
  muscle
10 Internal carotid artery
11 Pharynx
12 Hyoid bone (greater horn)
13 Trochlear nerve (IV)
14 Facial nerve (VII), vestibulocochlear nerve (VIII)
15 Glossopharyngeal nerve (IX)
16 Accessory nerve (XI)
17 Hypoglossal nerve (XII)
18 Vagus nerve (X), internal
   carotid artery
19 External carotid artery
20 Sympathetic trunk, superior
   sympathetic ganglion
21 Ansa cervicalis
22 Olfactory sulcus (termination)
23 Orbital gyri
24 Temporal lobe
25 Straight gyrus
26 Olfactory trigone, inferior temporal sulcus
27 Medial occipitotemporal gyrus
28 Parahippocampal gyrus, mamillary
   body, interpeduncular fossa
29 Pons and cerebral peduncle
30 Abducens nerve (VI)
31 Pyramid
32 Olive
33 Cervical spinal nerves
34 Cerebellum
35 Tonsil of cerebellum
36 Occipital pole
37 Longitudinal fissure
38 Olfactory bulb
39 Orbital sulcus of frontal lobe
40 Olfactory tract
41 Optic nerve (II) and anterior
   perforated substance
42 Optic chiasma, neural stalk
43 Optic tract
44 Oculomotor nerve (III)
45 Trigeminal nerve (V)
46 Facial nerve (VII)
47 Vestibulocochlear nerve (VIII)
48 Flocculus of cerebellum
49 Glossopharyngeal (IX) and
   vagus (X) nerves
50 Vermis of cerebellum

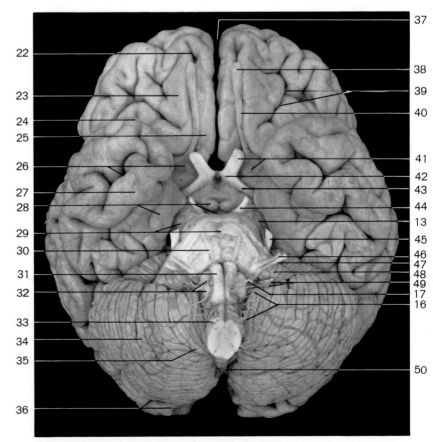

8.23 Brain stem with cranial nerves (posterior view). Cranial cavity opened and cerebellum removed.

8.24   Cranial nerves. Inferior view of the brain.

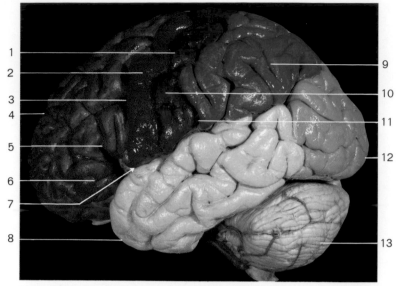

8.25 Brain, left hemisphere. Lateral view (frontal lobe to the left). Parts of the brain are color coded.

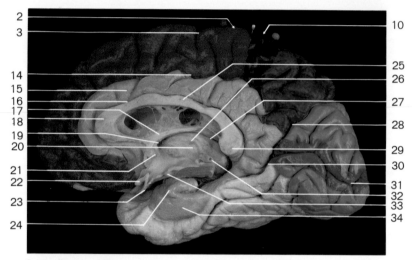

8.26 Brain, right hemisphere. Medial view (frontal lobe to the left). The midbrain has been cut and the cerebellum and brain stem removed.

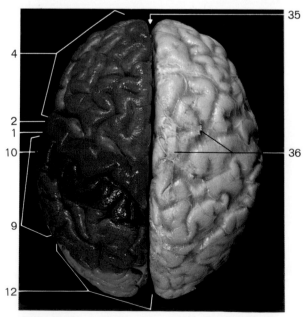

8.27 Brain, superior view. Right hemisphere is covered with pia mater. Note the arachnoid granulations.

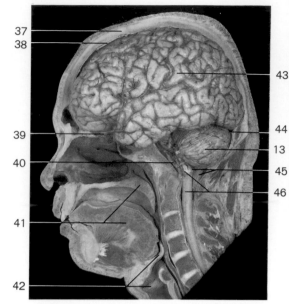

8.28 Dissection of the brain with pia mater *in situ*. The entire brain is shown. The facial part of the head has been cut through the midsagittal plane.

1 Central sulcus
2 Precentral gyrus (dark red)
3 Precentral sulcus
4 Frontal lobe (pink)
5 Ramus ascendens
6 Ramus anterior
7 Lateral sulcus
8 Temporal lobe (yellow)
9 Parietal lobe (blue)
10 Postcentral gyrus (dark blue)
11 Postcentral sulcus
12 Occipital lobe (green)
13 Cerebellum
14 Cingulate sulcus
15 Cingulate gyrus (light brown)
16 Sulcus of corpus callosum
17 Fornix
18 Genu of corpus callosum
19 Interventricular foramen
20 Intermediate mass of thalamus
21 Anterior commissure
22 Optic chiasma
23 Infundibulum (neural stalk)
24 Uncus
25 Trunk of corpus callosum
26 Thalamus, site of third ventricle
27 Stria medullaris
28 Parietooccipital sulcus
29 Splenium of corpus callosum
30 Communication of calcarine and parietooccipital sulcus
31 Calcarine sulcus
32 Epiphysis (Pineal gland)
33 Mamillary body
34 Parahippocampal gyrus (orange)
35 Longitudinal fissure
36 Arachnoid granulations
37 Calvaria and skin of the head
38 Dura mater (cut)
39 Olfactory bulb
40 Basilar artery
41 Soft palate, tongue
42 Epiglottis, vocal fold
43 Cerebrum
44 Tentorium cerebelli
45 Cerebellomedullary cistern
46 Medulla, spinal cord

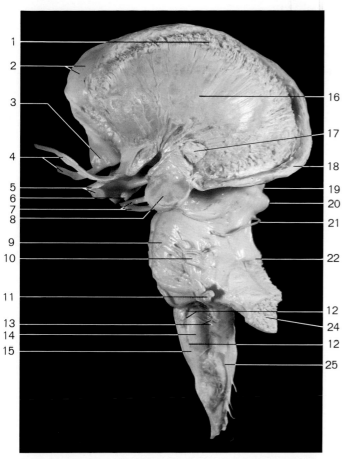

8.29 **Brain stem** (left lateral view). Cerebellar peduncles have been severed, cerebellum and cerebral cortex have been removed.

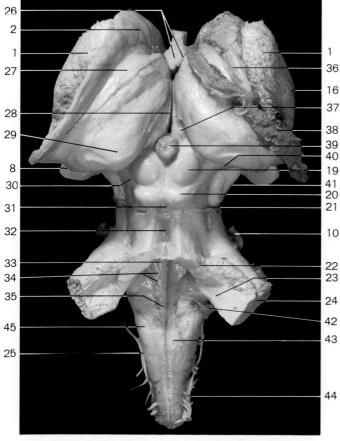

8.30 **Brain stem** (posterior view). Cerebellum removed.

1 Internal capsule
2 Head of caudate nucleus
3 Olfactory trigone
4 Olfactory tracts
5 Optic nerves
6 Infundibulum (neural stalk)
7 Oculomotor nerves
8 Amygdaloid body
9 Pons
10 Trigeminal nerve
11 Facial and vestibulocochlear nerves
12 Hypoglossal nerve
13 Glossopharyngeal and vagus nerves
14 Olive
15 Medulla oblongata
16 Lentiform nucleus
17 Anterior commissure
18 Tail of caudate nucleus
19 Superior colliculus
20 Inferior colliculus
21 Trochlear nerve
22 Superior cerebellar peduncle
23 Inferior cerebellar peduncle
24 Middle cerebellar peduncle
25 Accessory nerve
26 Columns of fornix (cut)
27 Lamina affixa
28 Third ventricle
29 Pulvinar of thalamus
30 Inferior brachium
31 Frenulum veli
32 Superior medullary velum
33 Facial colliculus
34 Striae medullares, rhomboid fossa
35 Hypoglossal triangle
36 Stria terminalis, thalamostriate vein
37 Habenular trigone
38 Choroid plexus of lateral ventricle
39 Epiphysis (Pineal gland)
40 Medial geniculate body
41 Cerebral peduncle
42 Choroid plexus of fourth ventricle
43 Clava of fasciculus gracilis
44 Dorsal root of cervical nerve
45 Cuneate tubercle

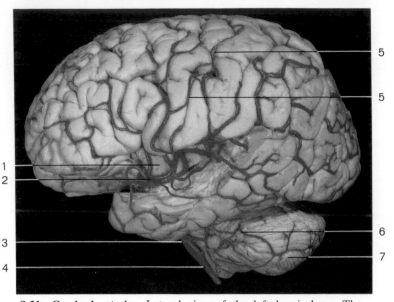

8.31   Cerebral arteries. Lateral view of the left hemisphere. The upper part of the temporal lobe has been removed to show the insula and cerebral arteries.

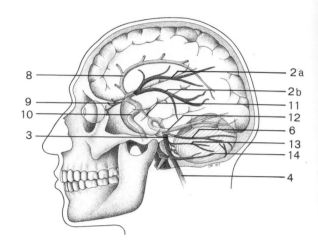

8.32   Arteries of the brain. Main branches of internal carotid and vertebral arteries (schematic drawing).

1	Insula	7	Cerebellum
2	Middle cerebral artery	8	Anterior cerebral artery
	a Parietal branches	9	Ophthalmic artery
	b Temporal branches	10	Internal carotid artery
3	Basilar artery	11	Posterior communicating artery
4	Vertebral artery	12	Posterior cerebral artery
5	Central sulcus	13	Anterior inferior cerebellar artery
6	Superior cerebellar artery	14	Posterior inferior cerebellar artery

15   Olfactory tract
16   Optic nerve
17   Infundibulum (neural stalk)
18   Oculomotor nerve, posterior communicating
       artery
19   Abducens nerve
20   Anterior spinal artery
21   Anterior communicating artery
22   Pons and superior cerebellar artery
23   Branches to pons
24   Medulla oblongata
25   Posterior spinal artery

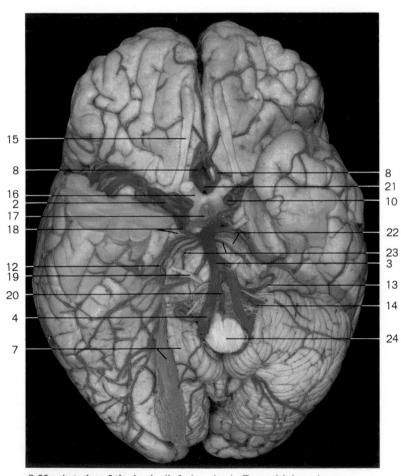

8.33   Arteries of the brain (inferior view). Frontal lobes above; right temporal lobe and cerebellum partly removed.

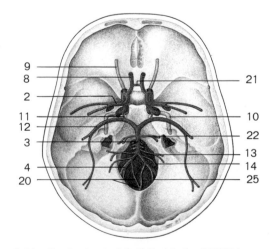

8.34   Cerebral arterial circle (circle of Willis). Superior view (schematic drawing).

1   Lateral longitudinal stria
2   Medial longitudinal stria
3   Genu of corpus callosum
4   Head of caudate nucleus
5   Septum pellucidum
6   Stria terminalis, thalamostriate vein
7   Thalamus
8   Choroid plexus of third ventricle
9   Choroid plexus of lateral ventricle
10   Splenium of corpus callosum
11   Posterior horn of lateral ventricle
12   Anterior horn of lateral ventricle
13   Interventricular foramen
14   Putamen of lentiform nucleus
15   Internal capsule
16   Inferior horn of lateral ventricle
17   Pes hippocampi
18   Fornix
19   Collateral eminence
20   Vermis of cerebellum with pia mater

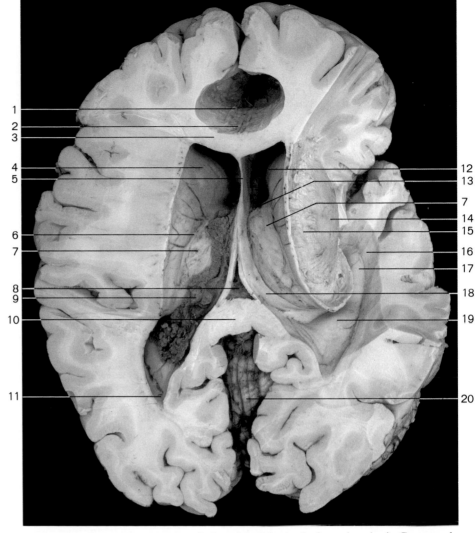

8.35   Lateral ventricle and subcortical nuclei of the brain (superior view). Corpus callosum has been partly removed. The lateral ventricle (at right) has been opened, and the insula with the claustrum, extreme and external capsules have been removed, to expose the lentiform nucleus and internal capsule.

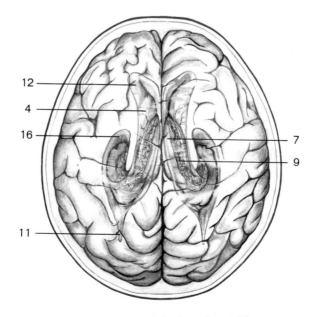

8.36   Superior view of the lateral ventricles (schematic drawing).

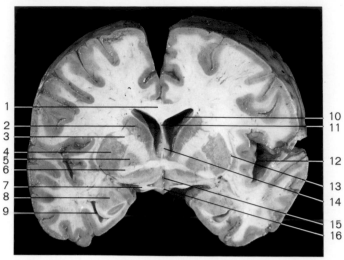

8.37   Frontal (coronal) section through the brain at the level of the anterior commissure.

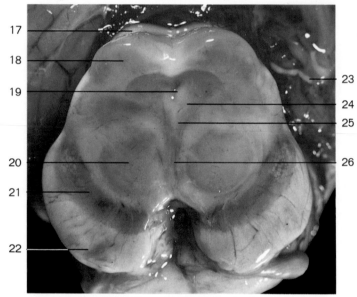

8.38   Cross section through the midbrain (mesencephalon) at the level of the superior colliculus. Anterior view.

1   Genu of corpus callosum
2   Head of caudate nucleus
3   Internal capsule
4   Putamen
5   Globus pallidus
6   Anterior commissure
7   Optic tract
8   Amygdaloid body
9   Inferior horn of lateral ventricle
10   Lateral ventricle
11   Septum pellucidum
12   Insula
13   External capsule
14   Column of fornix
15   Optic recess
16   Infundibulum
17   Inferior colliculus
18   Superior colliculus
19   Cerebral aqueduct (of Sylvius)
20   Red nucleus
21   Substantia nigra
22   Cerebral peduncle
23   Trochlear nerve
24   Gray matter
25   Nucleus of oculomotor nerve (nucleus of Edinger-Westphal)
26   Root of oculomotor nerve
27   Suprapineal recess
28   Pineal recess
29   Notch for posterior commissure
30   Posterior horns of right and left lateral ventricles
31   Fourth ventricle
32   Interventricular foramen (of Monro)
33   Anterior horn of lateral ventricle
34   Site of intermediate mass
35   Notch for anterior commissure
36   Third ventricle
37   Notch for optic chiasma
38   Infundibular recess
39   Lateral recess, lateral aperture (of Luschka)
40   Median aperture (of Magendie)
41   Cerebellomedullary cistern

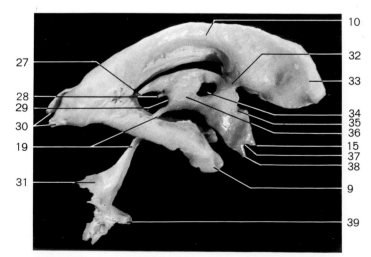

8.39   Cast of ventricular cavities of the brain. Lateral view, anterior horn to the right.

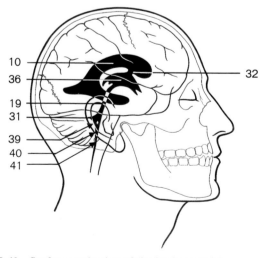

8.40   Surface projection of the brain ventricles (schematic drawing).

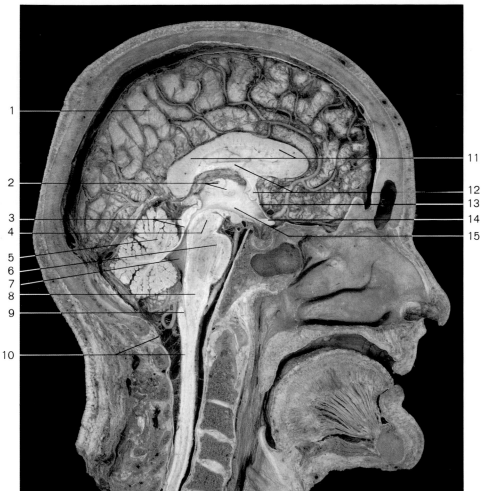

8.41 Midsagittal section through the head showing regions of the brain. Falx cerebri has been removed.

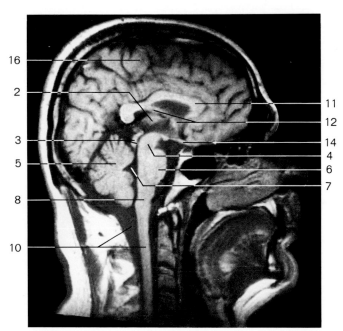

8.42 Midsagittal section through the head. Magnetic resonance image (MRI).

1 Left cerebral hemisphere
2 Thalamus, intermediate mass
3 Colliculi of midbrain, cerebral aqueduct
4 Cerebral peduncles
5 Cerebellum
6 Pons
7 Fourth ventricle
8 Medulla oblongata
9 Central canal
10 Cerebellomedullary cistern, spinal cord
11 Corpus callosum
12 Fornix, anterior commissure
13 Lamina terminalis
14 Optic chiasma
15 Hypothalamus
16 Cerebrum

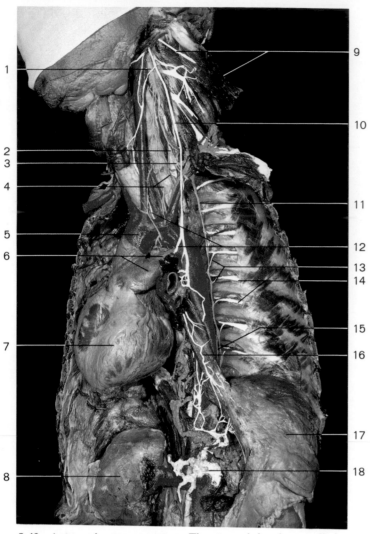

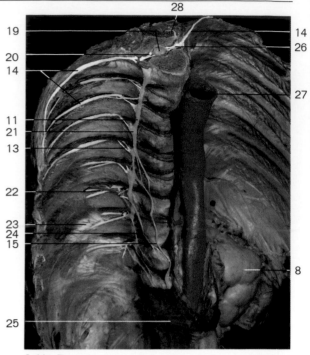

8.44   Posterior part of the thorax. Cross section at the level of the 5th thoracic segment showing the spinal nerves and connections to the sympathetic trunk.

8.43   Autonomic nervous system. The stomach has been pulled to the left. White = vagus nerve and spinal nerves; yellow = sympathetic nerves and ganglia; green = lymphatic trunks; red = arteries; blue = veins.

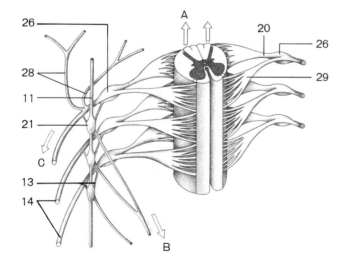

8.45   Organization of the spinal cord in structurally equal segments which form the paired spinal nerves (schematic drawing). A = connections to the brain; B = connections to the autonomic nervous system; C = connections to the trunk and extremities (intercostal nerves and plexus).

1   Superior cervical ganglion of sympathetic trunk
2   Medial cervical ganglion of sympathetic trunk
3   Inferior cervical ganglion (yellow) of sympathetic trunk
4   Cardiac nerves
5   Aortic arch
6   Ligamentum arteriosum, pulmonary trunk
7   Heart
8   Kidney
9   Accessory nerve, sternocleidomastoid
10   Vagus nerve
11   Sympathetic trunk
12   Left recurrent laryngeal nerves
13   Communicating rami
14   Intercostal nerves
15   Greater splanchnic nerve
16   Anterior vagal trunk, esophageal plexus
17   Stomach
18   Celiac ganglia
19   Spinal cord
20   Dorsal root of spinal nerve
21   Ganglion of sympathetic trunk
22   Intercostal artery and vein
23   Subcostalis muscle
24   Lesser splanchnic nerve
25   Inferior vena cava
26   Dorsal root ganglion
27   Aorta
28   Posterior branches of spinal nerve
29   Ventral root of spinal nerve

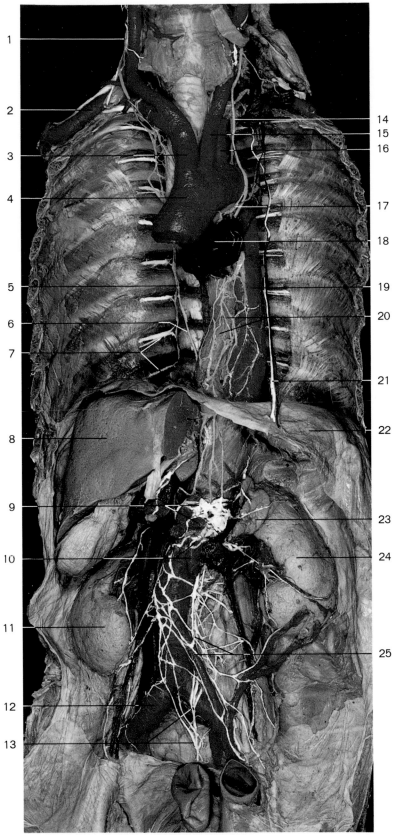

1 Right vagus nerve
2 Right axillary artery
3 Brachiocephalic trunk
4 Aortic arch
5 Sympathetic trunk
6 Greater splanchnic nerve
7 Intercostal nerves
8 Liver
9 Celiac trunk with celiac ganglion
10 Superior mesenteric ganglion and artery
11 Right kidney
12 Right common iliac artery
13 Superior hypogastric plexus and ganglion
14 Left vagus nerve
15 Left common carotid artery
16 Left subclavian artery
17 Left recurrent laryngeal nerve
18 Pulmonary trunk
19 Thoracic aorta
20 Esophageal plexus
21 Left phrenic nerve
22 Diaphragm
23 Left adrenal (suprarenal) gland
24 Left kidney
25 Inferior mesenteric ganglion and artery

8.46 Posterior wall of thoracic and abdominal cavities with sympathetic plexuses and ganglia (anterior view).
Red = arteries; blue = veins; yellow = parasympathetic nerves (vagus nerves); light green = sympathetic trunk and splanchnic nerves;
dark green = thoracic duct and lymph node;
white = spinal nerves and autonomic plexus;
orange = ligamentum arteriosum.

9. Sensory Organs of the Head

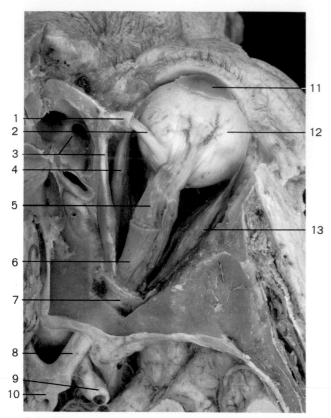

9.1 Right orbit with eyeball and extraocular muscles. Viewed from above. The roof of the orbit has been removed and the levator palpebrae superioris muscles has been severed.

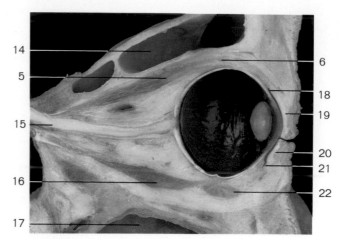

9.2 Sagittal section through the right orbit and eyeball (lateral view).

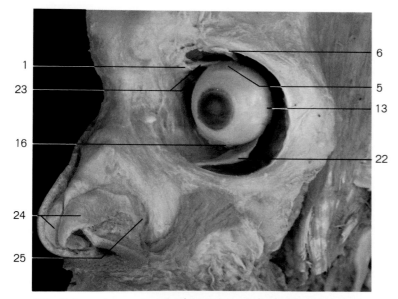

9.3 Extraocular eye muscles (anterolateral view). Medial rectus is behind the eyeball.

1 Trochlea of superior oblique muscle
2 Tendon of superior oblique muscle
3 Superior oblique muscle, ethmoid air cells
4 Medial rectus
5 Superior rectus
6 Levator palpebrae superioris
7 Common anular tendon
8 Optic nerve (intracranial part)
9 Internal carotid artery
10 Optic chiasma
11 Cornea
12 Eyeball
13 Lateral rectus
14 Frontal sinus
15 Optic nerve (extracranial part)
16 Inferior rectus
17 Maxillary sinus
18 Superior conjunctival fornix
19 Superior tarsus
20 Inferior tarsus
21 Inferior conjunctival fornix
22 Inferior oblique muscle
23 Superior oblique muscle and tendon
24 Greater alar cartilage
25 Lesser alar cartilage

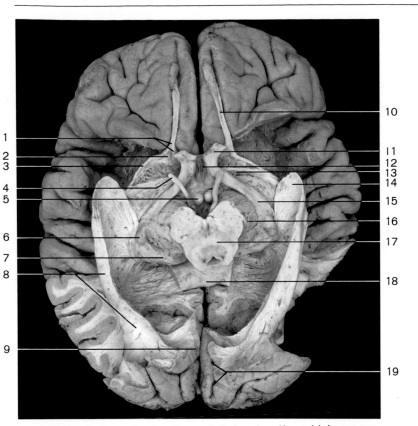

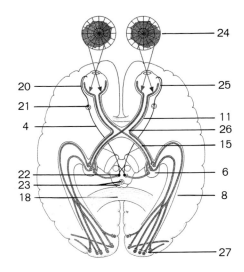

9.5 Diagram of the visual pathway and path of the light reflex.

9.4 Dissection of the visual pathway. Inferior view (frontal lobes at top, midbrain cut).

1 Medial olfactory stria	12 Infundibulum (neural stalk)	22 Accessory oculomotor nucleus	31 Inferior temporal arteriole and venule
2 Olfactory trigone	13 Anterior commissure	23 Colliculi of midbrain	32 Superior macular arteriole and venule
3 Lateral olfactory stria	14 Genu of optic radiation	24 Visual fields	33 Fovea centralis
4 Oculomotor nerve	15 Optic tract	25 Retina	34 Ciliary processes
5 Mamillary body	16 Trochlear nerve	26 Optic chiasma	35 Ora serrata
6 Lateral geniculate body	17 Midbrain	27 Visual cortex	36 Sclera
7 Pulvinar of thalamus	18 Corpus callosum	28 Superior temporal arteriole and venule	37 Choroid
8 Optic radiation	19 Calcarine sulcus	29 Optic disk	38 Lens (posterior view)
9 Cuneus of occipital lobe	20 Ciliary nerves (long and short)	30 Medial arteriole and venule	
10 Olfactory tract	21 Ciliary ganglion		
11 Optic nerve			

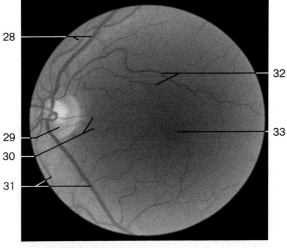

9.6 Fundus of a normal eye.

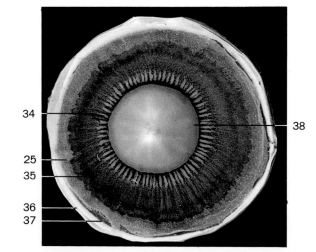

9.7 Anterior segment of the eyeball (posterior view). The opacity of the lens is an artifact.

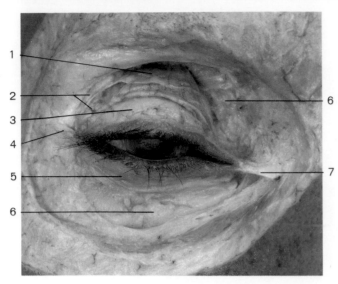

9.8A   Eyelids and lacrimal apparatus. Lacrimal gland and palpebrae are seen.

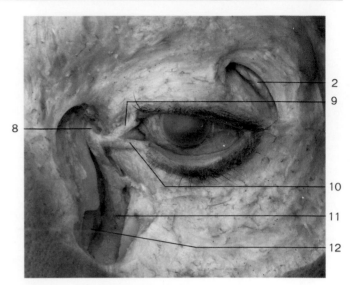

9.8B   Eyelids and lacrimal apparatus. Lacrimal duct, lacrimal sac and nasolacrimal duct are shown.

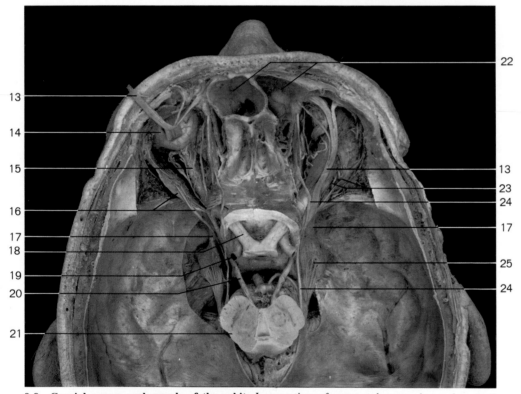

9.9   Cranial nerves and vessels of the orbit. Innervation of extraocular muscles and eyeball. The ciliary ganglion, located lateral to the optic nerve, cannot be observed in this field.

1 Levator palpebrae superioris	10 Inferior lacrimal canaliculus	19 Internal carotid artery (cut)
2 Lacrimal gland	11 Nasolacrimal duct	20 Oculomotor nerve
3 Superior tarsus	12 Mucous membrane of nasal cavity	21 Midbrain (cut)
4 Lateral palpebral ligament	13 Frontal nerve	22 Frontal sinus
5 Inferior tarsus	14 Eyeball	23 Lacrimal nerve and artery
6 Orbital fat	15 Short ciliary nerves	24 Trochlear nerve
7 Medial palpebral ligament	16 Abducens nerve, lateral rectus muscle	25 Trigeminal ganglion
8 Lacrimal sac	17 Ophthalmic nerve	
9 Superior lacrimal canaliculus	18 Optic nerve	

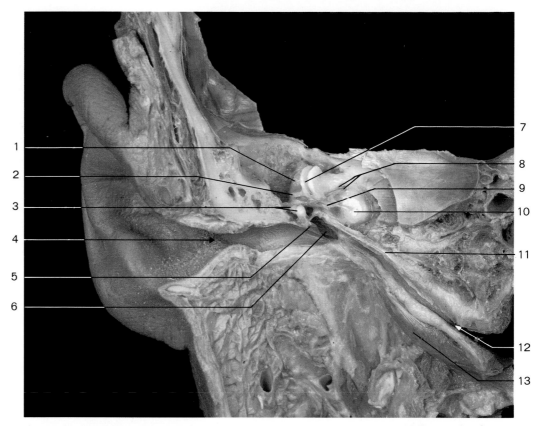

9.10 **Longitudinal section through the right external acoustic meatus and middle ear**, showing auditory ossicles and auditory tube.

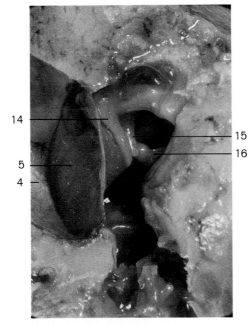

9.11 **Tympanic membrane and auditory ossicles.**

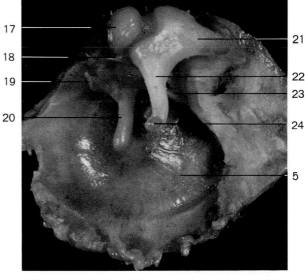

9.12 **Internal view of the tympanic membrane with malleus and incus.**

1	Posterior semicircular canal	9	Geniculate ganglion	17	Head of malleus
2	Lateral semicircular canal	10	Cochlea	18	Anterior ligament of malleus
3	Malleus, incus	11	Tensor tympani	19	Tendon of tensor tympani
4	External acoustic meatus	12	Auditory tube (pharyngeal orifice)	20	Handle of malleus
5	Tympanic membrane	13	Levator palatini	21	Short limb of incus
6	Tympanic cavity	14	Malleus	22	Long limb of incus
7	Anterior semicircular canal	15	Incus	23	Chorda tympani
8	Vestibulocochlear nerve, facial nerve	16	Stapes	24	Lentiform process

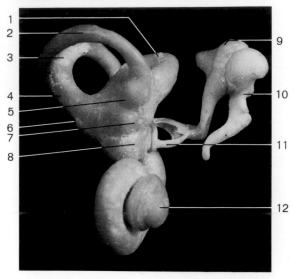

9.13   Chain of auditory ossicles and cast of bony labyrinth of the inner ear. Anterolateral view (left side).

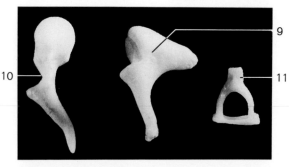

9.14   Auditory ossicles (isolated).

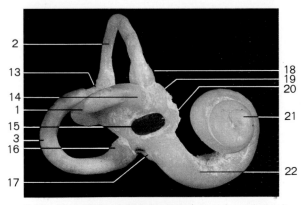

9.15   Cast of the right labyrinth (anterolateral view).

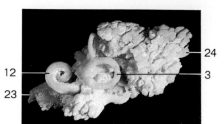

9.16   Cast of the labyrinth and mastoid cells (posterior view). Actual size.

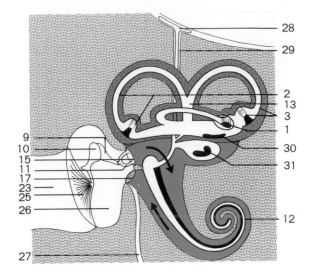

9.17   Auditory and vestibular apparatus (schematic drawing). Arrows indicate direction of sound waves. Blue=perilymphatic ducts.

 1   Lateral semicircular duct with ampulla
 2   Anterior semicircular duct with ampulla
 3   Posterior semicircular duct with ampulla
 4   Common limb
 5   Ampulla
 6   Beginning of endolymphatic duct
 7   Utricular prominens
 8   Saccular prominens
 9   Incus
10   Malleus
11   Stapes
12   Cochlea
13   Common limb
14   Lateral ampulla
15   Oval window for stapes
16   Posterior ampulla
17   Round window
18   Anterior ampulla
19   Elliptical recess
20   Spherical recess
21   Cupula of cochlea
22   Base of cochlea
23   External acoustic meatus
24   Mastoid air cells
25   Tympanic membrane
26   Tympanic cavity
27   Aqueduct of cochlea
28   Endolymphatic sac
29   Endolymphatic duct
30   Macula of utricle
31   Macula of saccule

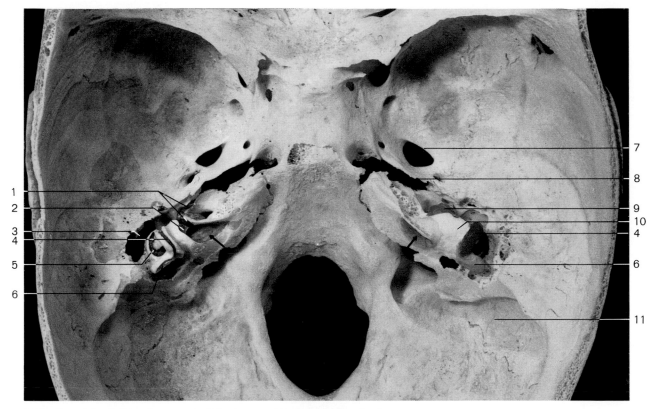

9.18 Bony labyrinth, petrous part of the temporal bone (superior view).
Semicircular canals are opened on the left side. Arrows=internal acoustic meatus.

1 Facial canal (blue), semicanalis of auditory tube
2 Superior vestibular area
3 Tympanic cavity
4 Anterior semicircular canal (red)
5 Lateral semicircular canal (green)
6 Posterior semicircular canal (yellow)
7 Foramen ovale
8 Foramen lacerum
9 Cochlea (orange)
10 Vestibule
11 Groove for sigmoid sinus
12 Precentral gyrus
13 Acoustic radiation
14 Lateral geniculate body
15 Optic tract and chiasma
16 Optic nerve
17 Oculomotor nerve
18 Postcentral gyrus
19 Optic radiation
20 Medial geniculate body
21 Tectal lamina
22 Trigeminal nerve
23 Cochlear part of vestibulocochlear nerve
24 Pyramidal tract

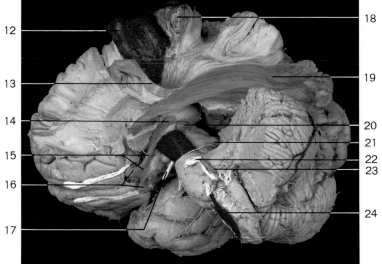

9.19 Visual and acoustic pathways in the brain. Left side.
Green = optic pathway; orange = acoustic pathway;
red = pyramidal tract.

10. Upper Respiratory and Digestive Organs in Head and Neck

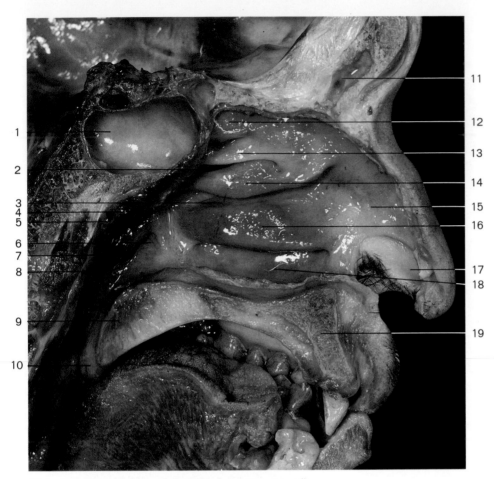

10.1 Lateral wall of the nasal cavity (septum removed).

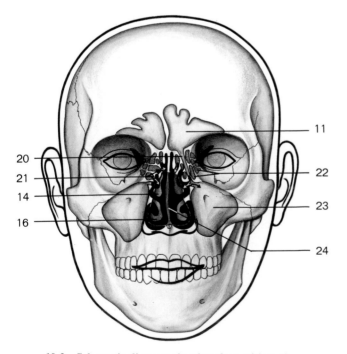

10.2 Schematic diagram showing the position of the paranasal sinuses. Arrows indicate openings.

1 Sphenoidal sinus
2 Superior meatus
3 Middle meatus
4 Tubal elevation
5 Pharyngeal tonsil
6 Pharyngeal orifice of auditory tube
7 Salpingopharyngeal fold
8 Pharyngeal recess
9 Soft palate
10 Uvula
11 Frontal sinus
12 Sphenoethmoidal recess
13 Superior nasal concha
14 Middle nasal concha
15 Atrium of middle nasal meatus
16 Inferior nasal concha
17 Vestibule of nasal cavity
18 Inferior meatus
19 Hard palate
20 Opening of frontal sinus
21 Opening of ethmoidal air cells
22 Ethmoidal air cells
23 Maxillary sinus
24 Nasal septum

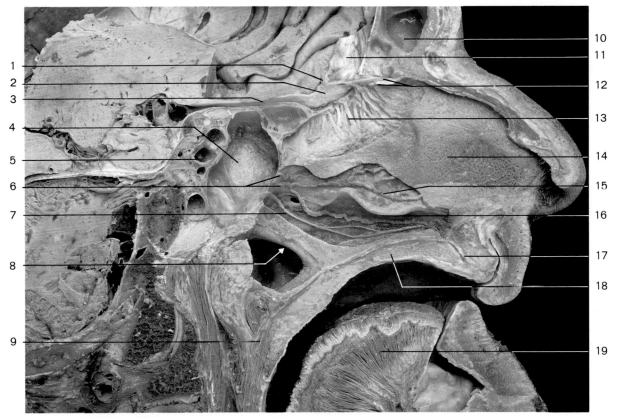

10.3 Nasal septum. Dissection of nerves and vessels.

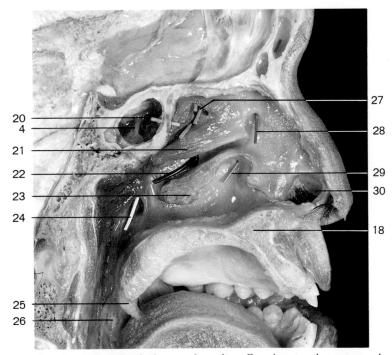

10.4 Lateral wall of the nasal cavity. Opening to the paranasal sinuses are indicated by probes. Inferior and middle nasal conchae have been partly removed.

1 Anterior ethmoidal artery
2 Olfactory bulb
3 Olfactory tract
4 Sphenoidal air sinus (relatively large)
5 Internal carotid artery
6 Posterior lateral nasal and septal arteries
7 Nasopalatine nerve
8 Choana
9 Soft palate
10 Frontal sinus
11 Crista galli of ethmoid
12 Nasal branch of anterior ethmoidal artery and nerve
13 Olfactory nerves
14 Nasal septum
15 Septal artery
16 Crest of nasal septum
17 Incisive canal
18 Hard palate
19 Tongue
20 Opening of sphenoidal sinus
21 Middle nasal concha
22 Opening of maxillary sinus (semilunar hiatus)
23 Inferior nasal concha
24 Pharyngeal opening of auditory tube
25 Uvula
26 Palatine tonsil
27 Openings of ethmoidal air cells
28 Opening of frontal sinus
29 Opening of nasolacrimal canal
30 Vestibule of nasal cavity

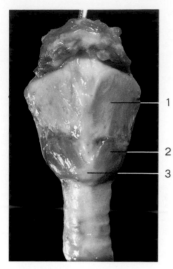

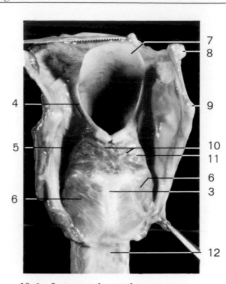

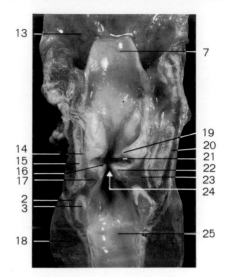

10.5   Laryngeal muscles.
Anterior view of the larynx.

10.6   Laryngeal muscles.
Posterior view of the larynx.

**10.7   Frontal (coronal) section
through larynx and trachea**
(posterior view).

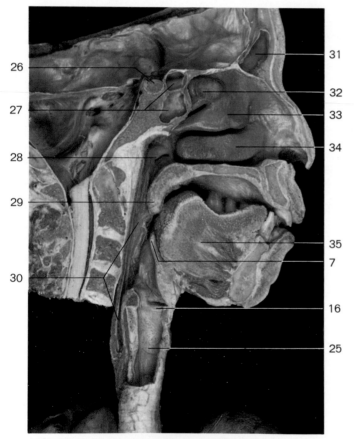

10.8   Sagittal section through nasal cavity and larynx.
The trachea has been partly opened.

1   Lamina of thyroid cartilage
2   Cricothyroid muscle
3   Cricoid cartilage
4   Aryepiglottic fold and muscle
5   Piriform fossa
6   Posterior cricoarytenoid muscle
7   Epiglottis
8   Greater horn of hyoid bone
9   Superior horn of thyroid cartilage
10 Oblique arytenoid muscle
11 Transverse arytenoid muscle
12 Membranous part of trachea
13 Root of tongue
14 Aryepiglotticus muscle
15 Thyroid cartilage
16 Vocal fold
17 Vocalis muscle
18 Thyroid gland
19 Vestibular fold
20 Thyrohyoid muscle
21 Ventricle
22 Vocal ligament
23 Lateral cricoarytenoid muscle
24 Rima glottidis
25 Trachea
26 Hypophysis (pituitary gland)
27 Sphenoidal sinus
28 Pharyngeal opening of auditory tube
29 Soft palate
30 Pharynx (oral and laryngeal parts)
31 Frontal sinus
32 Superior nasal concha
33 Middle nasal concha
34 Inferior nasal concha
35 Tongue

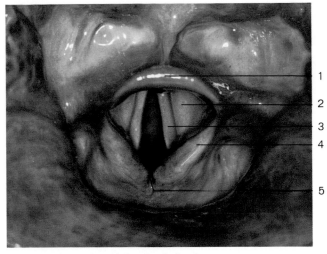

10.9   Superior view of the glottis *in vivo*.

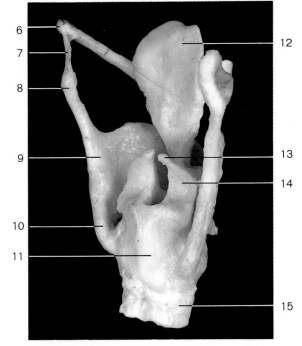

10.10   Cartilages of the larynx (oblique posterior view).

1	Epiglottis	
2	Vestibular fold	
3	Vocal fold	
4	Aryepiglottic fold	
5	Interarytenoid notch	
6	Greater horn of hyoid bone	
7	Thyrohyoid ligament	
8	Superior horn of thyroid cartilage	
9	Thyroid cartilage	
10	Inferior horn of thyroid cartilage	
11	Cricoid cartilage	

12	Epiglottic cartilage
13	Corniculate cartilage
14	Arytenoid cartilage
15	Trachea
16	Transverse arytenoid muscle
17	Posterior cricoarytenoid muscle
18	Membranous part of the trachea
19	Lamina of thyroid cartilage (cut)
20	Vocal ligament
21	Lateral cricoarytenoid muscle
22	Articular facet for thyroid cartilage (cricothyroid articulation)

23	Tracheal cartilages
24	Thyrohyoid membrane
25	Oblique arytenoid muscle
26	Lamina of cricoid cartilage
27	Fissure of glottis, vocal ligament
28	Vocalis muscle
29	Cricoarytenoid articulation
30	Arch of cricoid cartilage

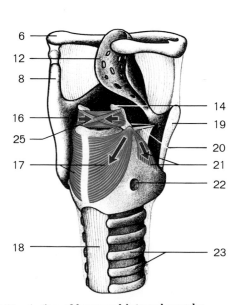

10.11   Action of laryngeal internal muscles (schematic drawing).

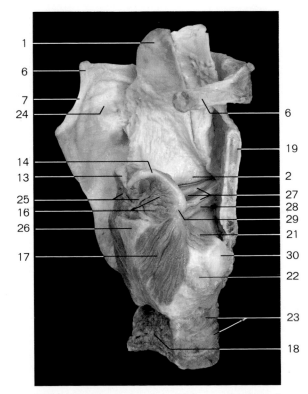

10.12   Laryngeal muscles (right posterolateral view). Half of thyroid cartilage has been removed.

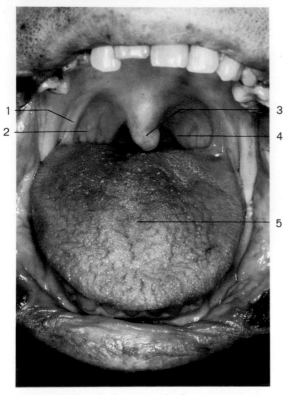

10.13 Oral cavity (anterior view).

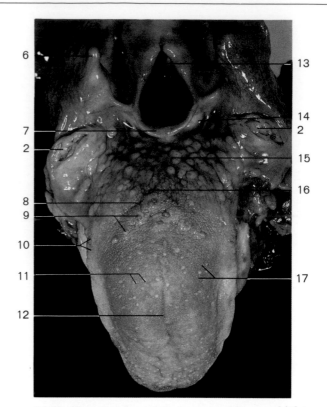

10.14 Dorsal surface of the tongue and laryngeal inlet.

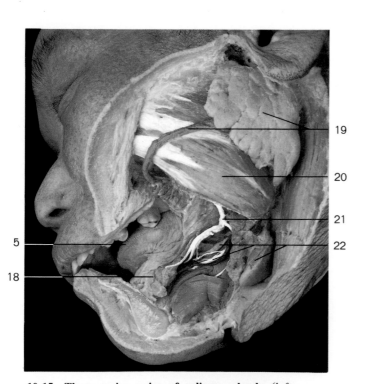

10.15 Three major pairs of salivary glands (left lateral view). Lateral wall of cheek and mandible have been removed. Green = salivary gland duct; white = nerve.

1 Palatoglossal arch
2 Palatine tonsil
3 Uvula
4 Palatopharyngeal arch
5 Tongue
6 Greater horn of hyoid bone
7 Epiglottis
8 Sulcus terminalis
9 Circumvallate papillae
10 Foliate papillae
11 Fungiform papillae
12 Median sulcus of tongue
13 Rima glottidis
14 Vallecula of epiglottis
15 Root of tongue, lingual tonsil
16 Foramen cecum
17 Filiform papillae
18 Sublingual gland
19 Parotid gland and duct
20 Masseter
21 Lingual nerve
22 Submandibular gland and duct,
   hypoglossal nerve

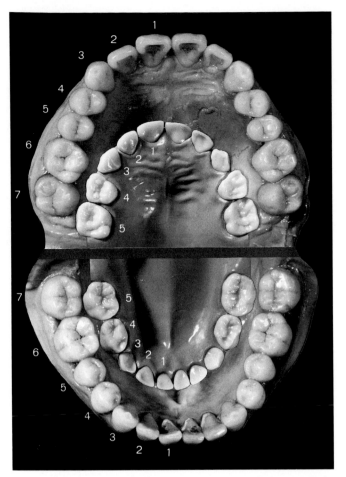

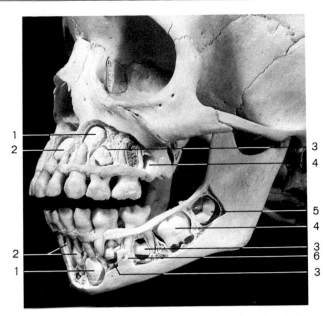

10.17 Deciduous teeth in child's skull. The developing crowns of the permanent teeth are shown in their sockets within the maxilla and mandible.

1 Permanent canine (cuspid)
2 Permanent incisors
3 Premolars (bicuspids)
4 First permanent molar
5 Second permanent molar
6 Mental foramen of mandible

10.16 Comparison of deciduous (inner rows) and permanent (outer rows) teeth. Notice that the alveolar arches of the maxillae (upper) and mandibles (lower), which hold deciduous teeth in the child and permanent teeth in the adult, have practically the same breadth.

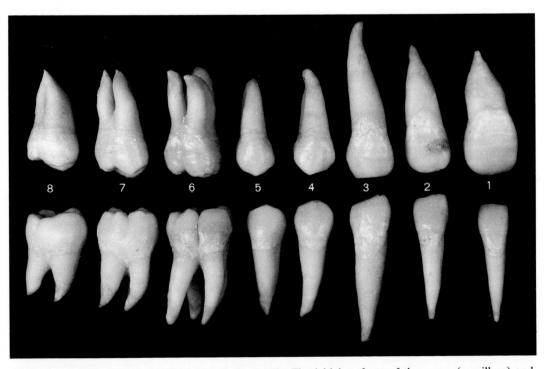

10.18 Isolated teeth from the right half of the mouth. The labial surfaces of the upper (maxillary) and lower (mandibular) teeth are shown. Digits indicate number of dental formula.

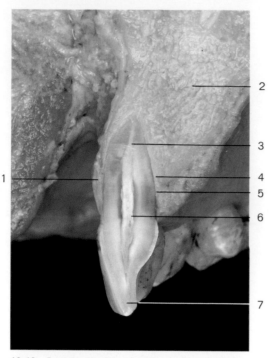

10.19   Longitudinal section through an incisor.

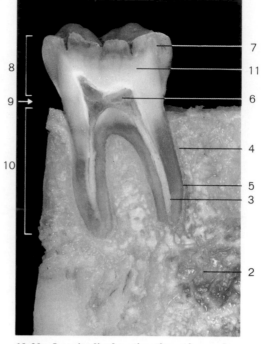

10.20   Longitudinal section through a molar.

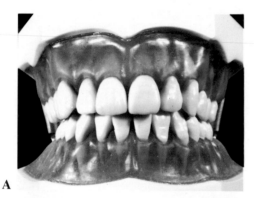

A

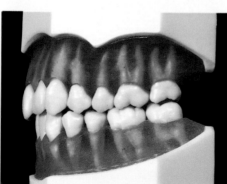

B

1   Gingiva
2   Alveolar bone
3   Root canal
4   Alveolar bone (cortical)
5   Cementum
6   Pulp
7   Enamel
8   Crown
9   Neck
10   Root
11   Dentin

10.21   Models of occlusion showing the roots of the teeth in the alveoli in anterior (A) and left lateral (B) views.

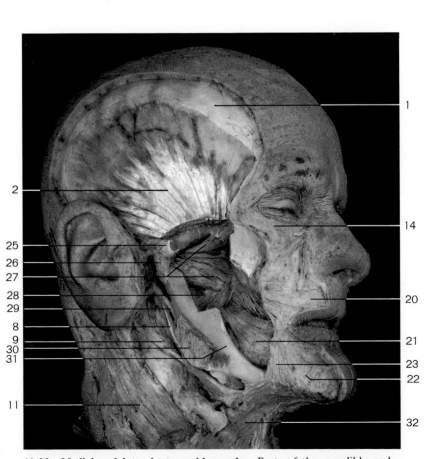

10.22 Temporalis and masseter muscles. The temporal fascia has been removed. The temporomandibular joint has been severed and the zygomatic arch is shown.

1   Galea aponeurotica
2   Temporalis
3   Occipital belly of occipitofrontalis
4   Temporomandibular joint
5   External acoustic meatus
6   Deep layer of masseter
7   Superficial layer of masseter
8   Stylohyoid
9   Posterior belly of digastric muscle
10 Internal jugular vein, external carotid artery
11 Sternocleidomastoid
12 Frontal belly of occipitofrontalis
13 Depressor supercilii
14 Orbicularis oculi
15 Transverse part of nasalis
16 Levator labii superioris alaeque nasi
17 Levator labii superioris
18 Levator anguli oris
19 Zygomaticus major
20 Orbicularis oris
21 Buccinator
22 Depressor labii inferioris
23 Depressor anguli oris
24 Submandibular gland
25 Zygomatic arch
26 Articular capsule of temporomandibular joint
27 Lateral pterygoid
28 Medial pterygoid
29 Styloglossus
30 Masseter (cut)
31 Mandible
32 Platysma
33 Head of mandible
34 Mylohyoid

10.23 Medial and lateral pterygoid muscles. Parts of the mandible and zygomatic arch have been removed to reveal the pterygoid region.

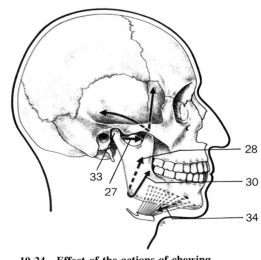

10.24 Effect of the actions of chewing muscles on the temporomandibular joint (schematic drawing).

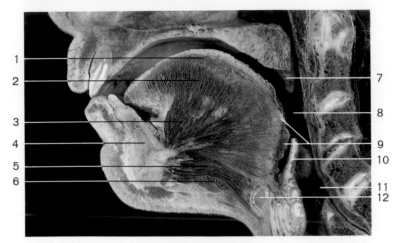

10.25   Sagittal section through the tongue.

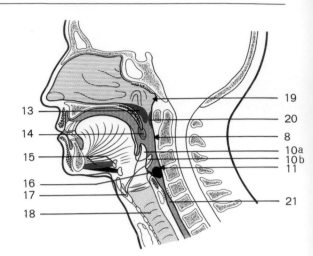

10.26   Schematic drawing showing the act of swallowing. The act of swallowing occurs in three stages: (1) the bolus is pushed backwards into the pharynx, the tongue is depressed with tip against the hard palate and pharyngeal muscles contract; (2) the bolus passes into the esophagus, the soft palate is raised to close off the internal nares, the larynx is elevated towards the epiglottis and vocal folds are adducted to cover the glottis; (3) the bolus is moved along by pharyngeal contractions and esophageal peristalsis.

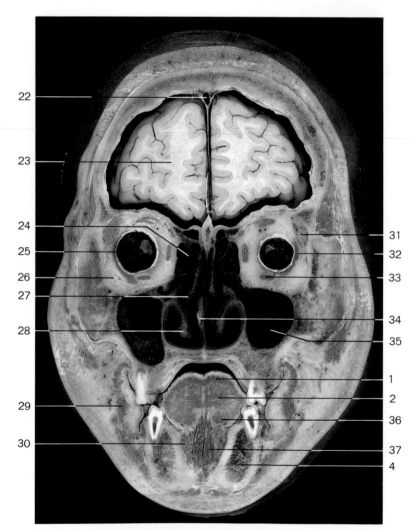

10.27   Frontal (coronal) section through the head.

1   Superior longitudinal muscle of tongue
2   Vertical and transverse muscles of tongue
3   Genioglossus, sublingual gland
4   Mandible
5   Geniohyoid
6   Mylohyoid
7   Uvula
8   Oropharynx
9   Lingual tonsil, root of tongue
10 Epiglottis
    a   Before swallowing
    b   During swallowing
11 Laryngopharynx
12 Hyoid bone
13 Soft palate
14 Palatine tonsil
15 Hyoid bone, geniohyoid muscle
16 Bolus
17 Thyroid cartilage of larynx
18 Trachea
19 Nasopharynx
20 Superior constrictor muscle of pharynx
21 Esophagus
22 Superior sagittal sinus
23 Frontal lobe of cerebrum
24 Ethmoidal labyrinth
25 Eyeball
26 Adipose body of orbit
27 Middle nasal concha
28 Inferior nasal concha
29 Buccinator
30 Sublingual gland
31 Lacrimal gland
32 Lateral rectus muscle
33 Inferior rectus muscle
34 Nasal septum
35 Maxillary sinus
36 Inferior longitudinal muscle of tongue
37 Genioglossus

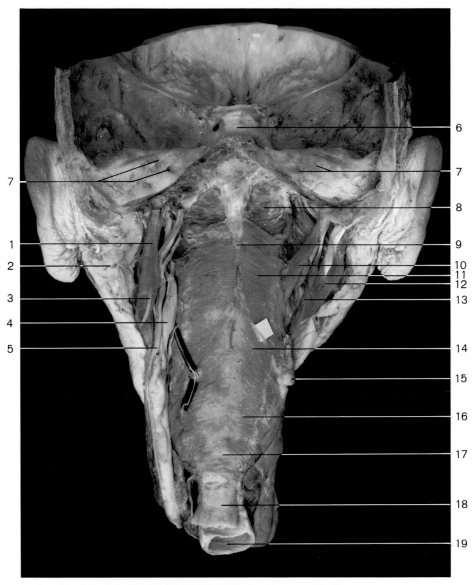

10.28 Muscles of the pharynx (posterior view).

1   Internal jugular vein
2   Parotid gland
3   Accessory nerve
4   Superior ganglion of sympathetic trunk
5   Vagus nerve
6   Sella turcica of sphenoid
7   Internal acoustic meatus, petrous part of temporal bone
8   Pharyngobasilar fascia
9   Fibrous raphe of pharynx
10 Stylopharyngeus
11 Superior constrictor of pharynx
12 Posterior belly of digastric muscle
13 Stylohyoid
14 Middle constrictor of pharynx
15 Greater horn of hyoid bone
16 Inferior constrictor of pharynx
17 Muscle-free area (Laimer's triangle)
18 Esophagus
19 Trachea
20 Medial pterygoid
21 Thyroid and parathyroid glands

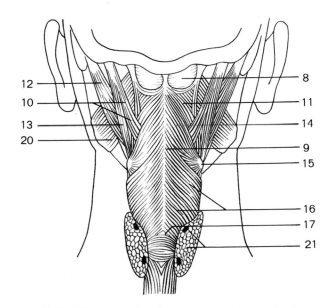

10.29 Muscles of the pharynx (schematic drawing).

11. Endocrine Organs

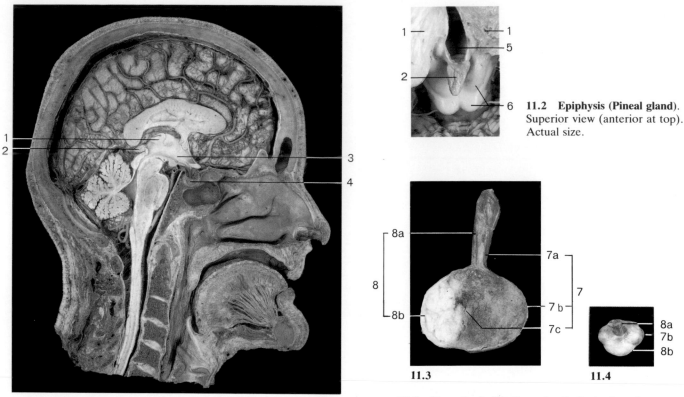

11.2 Epiphysis (Pineal gland). Superior view (anterior at top). Actual size.

11.3

11.4

11.1 Midsagittal section through the head showing the epiphysis (pineal gland) and hypophysis (pituitary gland), *in situ*.

11.3 Hypophysis (Pituitary gland). Sagittal section (anterior to right).
11.4 Hypophysis (Pituitary gland). Superior view (posterior at bottom). Actual size.

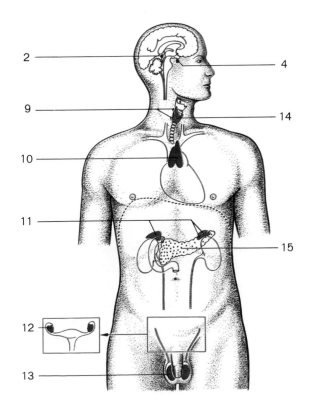

11.5 Endocrine organs in the human body (schematic drawing).

1 Thalamus
2 Epiphysis (pineal gland)
3 Hypothalamus
4 Hypophysis (pituitary gland)
5 Third ventricle
6 Tectal plates (colliculi of midbrain)
7 Adenohypophysis
   a pars tuberalis
   b Pars distalis of anterior lobe
   c Pars intermedia
8 Neurohypophysis
   a Infundibulum (neural stalk)
   b Pars nervosa of posterior lobe
9 Parathyroid glands
10 Thymus
11 Adrenal (suprarenal) glands
12 Ovaries
13 Testes
14 Thyroid gland
15 Pancreas (islets of Langerhans)

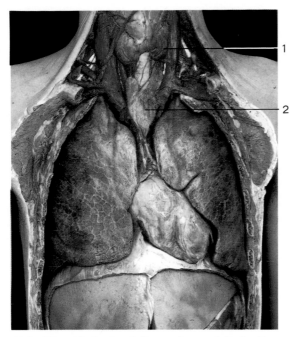

11.6 **Thyroid gland and thymus in an adult,** *in situ*.

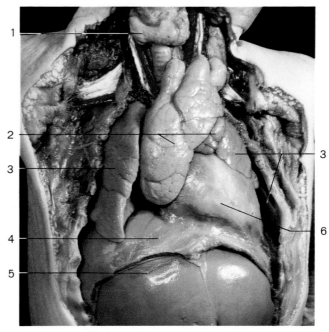

11.7 **Thyroid gland and thymus in a mature fetus,** *in situ*. Notice the large size of the fetal thymus compared to the small size of the atrophied adult thymus.

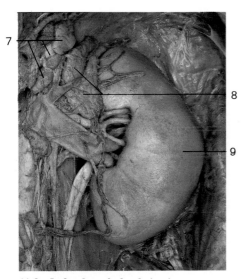

11.8 **Left adrenal gland,** *in situ* (anterior view).

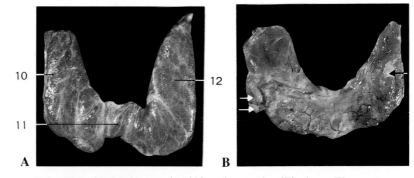

11.9 **Thyroid glands,** anterior (A) and posterior (B) views. The **parathyroid glands** (arrows) are on the posterior surface.

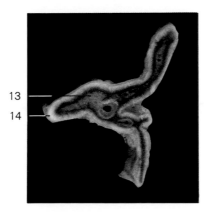

11.10 **Section of adrenal gland.**

1 Thyroid gland
2 Thymus
3 Lung
4 Diaphragm
5 Liver
6 Pericardial sac
7 Lymph nodes
8 Adrenal (suprarenal) gland
9 Kidney
10 Right lobe of thyroid
11 Isthmus
12 Left lobe of thyroid
13 Adrenal medulla
14 Adrenal cortex

12. Transverse Anatomy

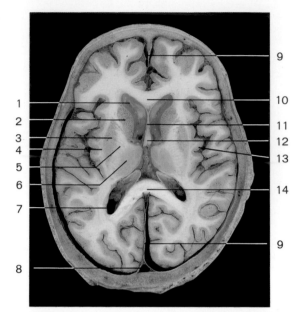

12.1A Transverse (horizontal) section through the head at the level of the basal ganglia and internal capsule.

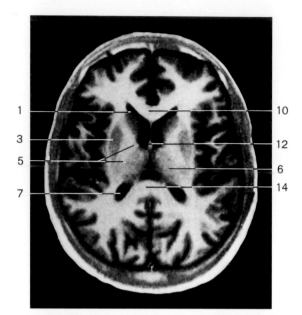

12.1B Magnetic resonance image (MRI) of the brain at corresponding level.

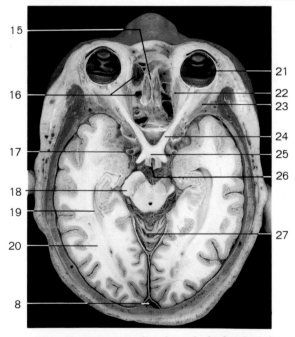

12.2A Transverse section through the head at the level of the eyeballs.

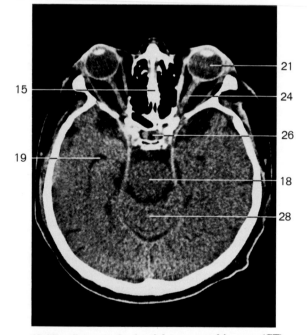

12.2B Computerized axial tomographic scan (CT) of the head at corresponding level.

1 Frontal (anterior) horn of lateral ventricle	9 Falx cerebri	19 Lateral ventricle
2 Head of caudate nucleus	10 Genu of corpus callosum	20 Cerebrum
3 Putamen	11 Septum pellucidum	21 Eyeball
4 Claustrum	12 Column of fornix	22 Medial rectus muscle
5 Internal capsule	13 Insula	23 Lateral rectus muscle
6 Thalamus	14 Splenium of corpus callosum	24 Optic nerve
7 Occipital (posterior) horn of lateral ventricle	15 Nasal septum	25 Optic chiasma
8 Superior sagittal sinus	16 Ethmoidal sinus	26 Infundibulum
	17 Internal carotid artery	27 Vermis of cerebellum
	18 Midbrain	28 Cerebellum

* CT and MRI scans supplied by courtesy of Siemens AG, Erlangen, FRG.

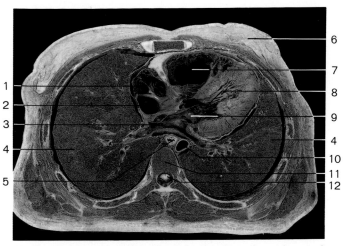

12.3A **Transverse section through the thoracic cavity at the level of the heart** (inferior view).

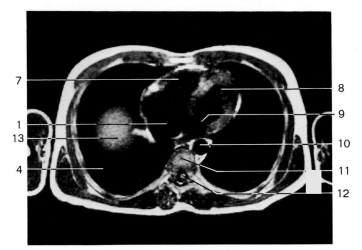

12.3B **MRI of the thorax at corresponding level.**

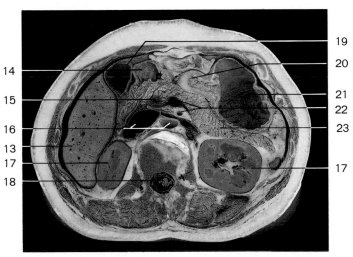

12.4A **Transverse section through the abdominal cavity at the level of the pancreas** (inferior view).

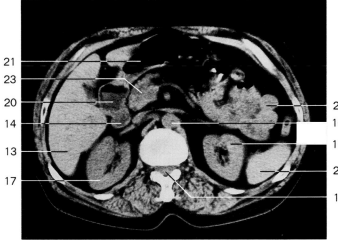

12.4B **CT of the abdomen at corresponding level.**

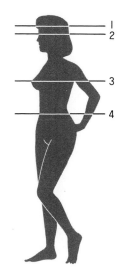

1	Right atrium	14	Duodenum
2	Superior vena cava	15	Common bile duct
3	Right pulmonary vein	16	Inferior vena cava, aorta
4	Lung	17	Kidney
5	Esophagus	18	Cauda equina of spinal cord, vertebral canal
6	Mammary gland	19	Right flexure of colon
7	Right ventricle	20	Pylorus
8	Left ventricle	21	Stomach
9	Left atrium	22	Superior mesenteric artery and vein
10	Aorta (thoracic part)	23	Pancreas
11	Thoracic vertebra	24	Small intestine
12	Spinal cord	25	Spleen
13	Liver		

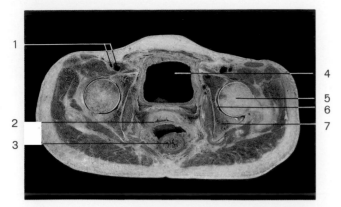

12.5A Transverse section through female pelvic cavity at the level of the head of femur (inferior view).

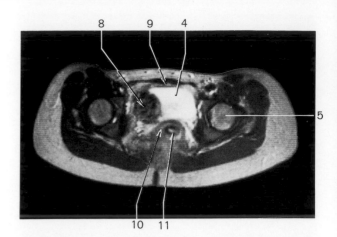

12.5B MRI of female pelvis at corresponding level.

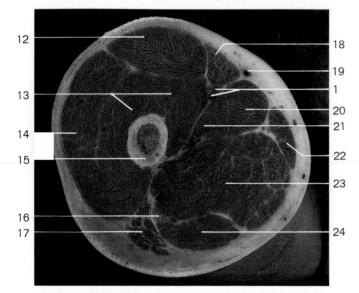

12.6A Transverse section through proximal part of right thigh (inferior view).

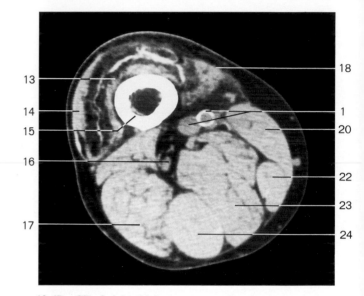

12.6B CT of right thigh at corresponding level.

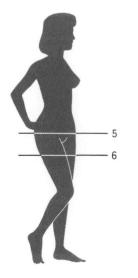

1	Femoral artery and vein	13	Vastus intermedius and medialis
2	Vagina	14	Vastus lateralis
3	Rectum	15	Femur
4	Urinary bladder	16	Sciatic nerve
5	Head of femur	17	Gluteus maximus
6	Margin of acetabulum	18	Sartorius
7	Part of hip bone	19	Saphenous vein
8	Intestine	20	Adductor longus
9	Wall of urinary bladder	21	Adductor brevis
10	Myometrium of uterus	22	Gracilis
11	Endometrium of uterus	23	Adductor magnus
12	Rectus femoris	24	Semitendinosus

INDEX

Reference page numbers to figures in Chapters 1 through 12 are in Roman. Reference page numbers for the text in Chapter 1 are in *italics*. Page numbers for tables are followed by *t*.